AF805354

Springer Geophysics

The Springer Geophysics series seeks to publish a broad portfolio of scientific books, aiming at researchers, students, and everyone interested in geophysics. The series includes peer-reviewed monographs, edited volumes, textbooks, and conference proceedings. It covers the entire research area including, but not limited to, geodesy, planetology, geodynamics, geomagnetism, paleomagnetism, seismology, and tectonophysics.

More information about this series at http://www.springer.com/series/10173

Xiaoqin Cui · Laurence Lines
Edward Stephen Krebes · Suping Peng

Seismic Forward Modeling of Fractures and Fractured Medium Inversion

Springer

Xiaoqin Cui
Department of Geoscience
University of Calgary
Calgary, AB
Canada

Edward Stephen Krebes
Department of Geoscience
University of Calgary
Calgary, AB
Canada

Laurence Lines
Department of Geoscience
University of Calgary
Calgary, AB
Canada

Suping Peng
China University of Mining and Technology (Beijing)
Beijing
China

Springer Geophysics
ISBN 978-981-10-3583-8 ISBN 978-981-10-3584-5 (eBook)
DOI 10.1007/978-981-10-3584-5

Library of Congress Control Number: 2017940353

© Springer Nature Singapore Pte Ltd. 2018
This work is subject to copyright. All rights are reserved by the Publisher, whether the whole or part of the material is concerned, specifically the rights of translation, reprinting, reuse of illustrations, recitation, broadcasting, reproduction on microfilms or in any other physical way, and transmission or information storage and retrieval, electronic adaptation, computer software, or by similar or dissimilar methodology now known or hereafter developed.
The use of general descriptive names, registered names, trademarks, service marks, etc. in this publication does not imply, even in the absence of a specific statement, that such names are exempt from the relevant protective laws and regulations and therefore free for general use.
The publisher, the authors and the editors are safe to assume that the advice and information in this book are believed to be true and accurate at the date of publication. Neither the publisher nor the authors or the editors give a warranty, express or implied, with respect to the material contained herein or for any errors or omissions that may have been made. The publisher remains neutral with regard to jurisdictional claims in published maps and institutional affiliations.

Printed on acid-free paper

This Springer imprint is published by Springer Nature
The registered company is Springer Nature Singapore Pte Ltd.
The registered company address is: 152 Beach Road, #21-01/04 Gateway East, Singapore 189721, Singapore

This book is dedicated to Xiaoqin's husband Yuanle Sun, and daughter Yi Sun.

Preface

The goal of this book is to develop a methodology for enhancing fracture detection and correctly delineating reservoirs with fractures. From the rock mechanics to seismic forward modeling then AVO inversion, the book deeply explores the mechanical formation of fractures and fractured media and presents an enhanced fracture detection technique that uses a new finite-difference scheme to accurately model fractures and analyze the fracture response in seismic traveltime and amplitude, and then develops a method for accurate reservoir delineation by deriving new AVO fracture equations to correctly estimate the properties of the fractured medium, the host medium, and the fractured medium with impedance contrast.

With the long wavelength assumption, a linear slip interface is equivalent to a fracture interface that satisfies the nonwelded contact boundary conditions. Therefore, the fractured medium can be regarded as a combination of a fracture, or a set of fractures, and a host medium as a horizontally fractured medium is composed of a horizontal fracture embedded into a homogeneous isotropic host medium; and a vertically fractured medium is formed by inserting a vertical fracture into a homogeneous isotropic host medium; an orthogonally fractured medium is assembled from a vertical fracture and a homogeneous VTI host medium, or a horizontal fracture and a homogeneous HTI host medium, or two orthorhombic fractures and a homogeneous isotropic host medium. Theoretically, the composed fractured media are effectively equal to the transversely isotropic medium with a symmetric axis (TI) media.

New finite-difference schemes with imposed explicit boundary conditions for horizontal, vertical, and orthorhombic fractures are implemented to generate seismograms that precisely illustrate the fracture representations in seismic data. The results indicate that the fractures are detectable, even though the fractured medium does not have impedance contrasts, and that the fractured medium can be characterized as a transversely isotropic medium. An analysis of how fractures are represented in seismic data can help in fracture detection in geoscience.

New exact equations for the reflection and transmission coefficients of a fractured medium with impedance contrast and new approximate AVO equations including fracture parameters are derived. This provides a way to separate the surface seismic records into the fracture reflections and reflections of the impedance contrast. Therefore, the fracture, the host medium, and the fractured medium with impedance contrast properties can be estimated from seismic data to correctly delineate the reservoir characterization.

Calgary, Canada — Xiaoqin Cui
Calgary, Canada — Laurence Lines
Calgary, Canada — Edward Stephen Krebes
Beijing, China — Suping Peng

Acknowledgements

This book has come from a Ph.D. thesis in geoscience at the University of Calgary, which was completed in September 2015.

We would like to thank the members of the Ph.D. committee, Dr. Gary F. Margrave, Dr. Chris Clarkson, and Dr. Brij Maini for their help in research. We would also like to express our appreciation to all CREWES and CHORUS staff and students for their help and support. We thank Joan Embleton for editing with the English in the thesis, Peter Manning for his FD staggered grid discussion and Tianci Cui for her help with the MATLAB displaying. Special thanks also go to Peng Cheng, Zaiming Jiang, Faranak Mahmoudian and Hassan Khaniani for their help and useful discussions.

Many thanks also go to the China University of Mining and Technology for permission and use of the real seismic dataset. We also thank Dr. Wenfeng Du in the China University of Mining and Technology for her support and friendship.

Contents

Symbols and Abbreviations

$\sigma, \sigma_{i,j}$	Stress and stress tensor component $i,j = 1, 2, 3$
$\varepsilon, \varepsilon_{i,j}$	Strain and strain tensor component $i,j = 1, 2, 3$
F	Force
S	An area
Δu_i	Displacement change
Δl_i	Length change
Φ_i	Angle of the distortion shape
C_{ijkl}	Rock stiffness $k, l = 1, 2, 3$
δ_{ij}	Kronecker delta
λ	Rock property: Lambda
μ	Rock property: Mu
$\mathbf{S}$	Rock compliances
S_{N}	Normal compliance of fracture
S_{T}	Tangential compliance of fracture
E_{N}	Dimensionless normal compliance of fracture
E_{T}	Dimensionless tangential compliance of fracture
Δ_{N}	Dimensionless normal weakness of fracture
Δ_{T}	Dimensionless tangential weakness of fracture
R, R'	Reflection coefficients
T, T'	Transmission coefficients
$C_{\mathrm{H\_iso}}$	Stiffness of the isotropic host medium
G_{VTI}	Abelian group of VTI medium
$G_{\mathrm{H\_iso}}$	Abelian group of isotropic host medium
$G_{\mathrm{h\_f}}$	Abelian group of horizontal fracture
C_{VTI}	Stiffness of the VTI medium
$S_{\mathrm{v\_f}}$	Compliance of the vertical fracture
G_{HTI}	Abelian group of HTI medium
$G_{\mathrm{v\_f}}$	Abelian group of vertical fracture
C_{HTI}	Stiffness of HTI medium
G_{orth}	Abelian group of orthorhombic medium

$G_{\mathrm{H\_VTI}}$	Abelian group of VTI host medium
C_{orth}	Stiffness of orthorhombic medium
$\boldsymbol{u}^{+}$	Displacements at positive side of the interface
$\boldsymbol{u}^{-}$	Displacements at negative side of the interface
$\boldsymbol{\sigma}^{+}$	Stresses at positive side of the interface
$\boldsymbol{\sigma}^{-}$	Stresses at negative side of the interface
$\boldsymbol{s}.\boldsymbol{x}$	Direction of wave
$\boldsymbol{d}$	Polarization of wave
$\mathrm{P}_1^{\backprime}$	Amplitude of P-wave incidence
P_1'	Amplitude of P-wave reflection
$\mathrm{P}_2^{\backprime}$	Amplitude of P-wave transmission
S_1'	Amplitude of PS-wave reflection
$\mathrm{S}_2^{\backprime}$	Amplitude of PS-wave transmission
θ_1	Angle of incident P-wave
θ_2	Angle of transmission P-wave
ϑ_1	Angle of reflection PS-wave
ϑ_2	Angle of transmission PS-wave
α_n	P-wave velocity in n medium $n = 1, 2$
β_n	S-wave velocity in n medium
ρ_n	Density in n medium
θ_m	Incident angle of plane wave at layer m
P	Ray parameter
R_{w}	Reflection coefficients at welded interface
T_{w}	Transmission coefficients at welded interface
ω	Frequency of incident plane wave
Z_n	Impedance of the media
X, Z	x-normal boundary, z-normal boundary
x, z	Grid indexes
$\Delta x, \Delta z$	Step size in x-direction and z-direction
$\tilde{\boldsymbol{u}}$	Fictitious displacement
$\gamma_{\mathrm{Th}}, \varepsilon_{\mathrm{Th}}, \delta_{\mathrm{Th}}$	Thomsen's anisotropy parameters
$\gamma, \varepsilon, \delta$	Anisotropy parameters in VTI medium
$\gamma^{v}, \varepsilon^{v}, \delta^{v}$	Anisotropy parameters in HTI medium
$\gamma^{(1)}, \varepsilon^{(1)}, \delta^{(1)}$	Anisotropy parameters in a vertical symmetric plane normal to x_1-axis in orthorhombic medium
$\gamma^{(2)}, \varepsilon^{(2)}, \delta^{(2)}$	Anisotropy parameters in a horizontal symmetric plane normal to x_2-axis in orthorhombic medium
$\delta^{(3)}$	Anisotropy parameters in a vertical symmetric plane normal to x_3-axis in orthorhombic medium
a	Courant condition
Δd	Space interval
Δt	Time interval
$S(t)$	Seismic wavelet in time domain
$W(t)$	Source wavelet

$R(t)$	Reflectivity serious
$f(t)$	Inverse source wavelet
PS_1	Fast converted shear wave
PS_2	Slow converted shear wave
R	Radial component
T	Transverse component
$\mathcal{R}$	Rotator
φ	Azimuth
$R_{w}(\theta)$	Reflection coefficients under welded boundary conditions
$R_{non\_w}(\theta)$	Reflection coefficients under nonwelded boundary conditions
r_{α}	Reflectivity of P-wave velocity
r_{β}	Reflectivity of S-wave velocity
r_{ρ}	Reflectivity of density
ACP	Asymptotic convert point
AGC	Automatic gain correction
AVO/AVAZ	Amplitude versus offset/azimuth
CCP	Common conversion point (CCP)
CMP	Common middle point
COCA	Common offset common azimuth
HTI	Transverse isotropy medium with horizontal symmetry axis
NMO	Normal move out
VTI	Transverse isotropy medium with vertical symmetry axis

List of Figures

List of Tables

Chapter 1
Introduction

Abstract Geological fractures play an important role in hydrocarbon exploration and recovery. In order to enhance the geophysical prediction of the fracture distribution, and to accurately estimate all of the medium properties related to fractures in the reservoir, the fracture characterizations and the fracture representations in seismic data are needed to be understood inevitably. Therefore, the fracture models, the fracture assumptions, and the boundary conditions that constrain seismic waves at the fracture are studied in this Chapter. Finite-difference forward modeling methods are described to precisely model fractures, with the resulting synthetic seismograms being modified by the fracture features in terms of traveltime and amplitude. AVO (amplitude variation with offset) methods are reviewed for the accurate prediction of all rock properties that are related to the fractures, for the purpose of correctly delineating the reservoir. Also, the structure of this book is presented in the last section.

Keywords Geological fractures · Fracture parameters · Fracture models · Fracture assumptions · Type of fracture media · Finite-difference · Forward modeling · AVO inversion

1.1 Motivation and Objectives

Fracture characterization is of great practical importance in hydrocarbon exploration and recovery because it is one of the main factors for determining reservoir parameters, and for controlling well-drilling planning and seismic data imaging quality. First, fractures can increase reservoir permeability and change the fluid flow rate in a tight reservoir, which is important because large portions of the hydrocarbons in the world are trapped in tight reservoirs (Nelson 1985). Second, a reverse effect is that the fractures can create a "thief zone" in the caprock by providing paths for the injected steam to escape during hot production (e.g. in a SAGD project) and thereby cause a decrease or termination of hydrocarbon production (Massonnat 1994). Third, the fractal network of wormholes in CHOPS needs to be located for

© Springer Nature Singapore Pte Ltd. 2018

X. Cui et al., *Seismic Forward Modeling of Fractures and Fractured Medium Inversion*, Springer Geophysics, DOI 10.1007/978-981-10-3584-5_1

optimal placement of infill wells (Yuan et al. 1999; Lines 2003, 2007, 2008), while the wormholes can be simulated by a stacked fracture network which produces abnormal amplitudes in seismic data (Cui 2013b). Fourth, fracture-induced anisotropy affects seismic imaging quality because fracture-induced anisotropy has azimuth-dependent characteristics. Especially, at the fracture interface, the splitting of fast and slow shear waves of converted PS data with time lags cause a low signal-to-noise ratio (S/N) data or an incorrect image (Crampin 1985; Lawton 1999; Bakulin 2000a, b, c; Cary 2008; Bale 2009; Thomsen 1995). Therefore, accurate location of fracture distributions is a task of top priority in the oil and gas industry.

Geophysical data have the advantages of lower acquisition costs, wider coverage, and deeper penetration for obtaining global geological information. However, the detection of subsurface fractures in geophysics is a challenging problem. Besides the low seismic resolution and the complexities of geological bodies in the subsurface, the main contributing causes for the problems are the challenge of understanding the response of fracture properties in seismic data and the correct estimation of all properties from seismic data.

Therefore, the motivation of this book is to meet the imminent industrial demands as described above, i.e., to deeply understand fracture representations in seismic data in order to enhance the geophysical prediction of the fracture distribution and to accurately estimate all of the medium properties related to fractures for the correct delineation of the reservoir. This ultimately benefits reservoir and mining engineers, geologists and geophysicists with respect to optimizing the hydrocarbon reservoir, well performance and safety mining.

The main objective of this book is to develop a methodology for enhancing fracture detection and optimizing the delineation of the reservoir. The emphasis of the book is on the study of fractures and fractured medium formations, and on knowledge of the fracture model and the boundary conditions that constrain seismic waves at the fracture. New forward modeling schemes are proposed to precisely model fractures, with the resulting synthetic seismograms being modified by the fracture features in terms of traveltime and amplitude, in order to improve fracture detection. Also, new AVO equations are developed for the accurate prediction of all rock properties that are related to the fractures, for the purpose of correctly delineating the reservoir.

1.2 Geological Fractures

Geological fractures are pairs of distinctly separated surfaces in the formation which are related to permanent rock crack deformation (Jaeger 1969; Price 1966; Priest and Hudson 1976; Schultz and Fossen 2008). Crack formation in a rock occurs when the rock strain is over a certain threshold and the rock loses cohesion since it is continuously subjected to stress.

Fundamentally, fractures are described in terms of height, length, density, orientation, and the opening size. The fracture height and length are measurements of the spatial extent of the fracture in the medium, and are governed primarily by in situ stress and the related rock properties (Dunphy and David 2011). The fracture density measures the number of fractures in units per meter in a certain direction and is related to reservoir permeability. Commonly, the higher the fracture density, the higher the permeability if the different fractures conduct fluids in the reservoir (Singhal and Gupta 1999). The fracture orientation is based on the relationship of three local principal compressive stresses and the fracture orientation is parallel to the direction of maximum compressive stress and perpendicular to the direction of minimum compressive stress (Anderson 1905, 1951). Thus, most fractures in the reservoir are vertical or nearly vertical with several azimuths because the compression due to overburden deposits globally dominates the stress, while the minimum stress locally varies depending on the tectonic movement in the reservoir. The fracture openings, or the fracture aperture, is a small displacement at the fracture plane. The fracture opening may remain unfilled or may get subsequently filled by secondary minerals or some fluids (Aguilera 1998; Romm 1985). The fracture opening is a function of the depth, and is much shorter than the seismic wavelength, hence, the fracture opening parameter is generally neglected from a seismic viewpoint in geophysical fracture studies.

Commonly, geological outcrops provide near-surface information (Gary 2008), but such information is difficult to accurately interpolate in the deeper areas. Petrophysical well logs are regarded as the most reliable data because they directly acquire information from the reservoir with a high resolution (Rider 2002; Schlumberger 1989), however, well logs are too sparsely located. While seismic data with wider coverage and deeper penetration have a big potential for fracture detection (Yu and Telford 1973), even though they pose a big challenge due to the lack of knowledge of the fracture response in seismic data. Therefore, an integrated method, involving geological, petrophysical, and geophysical technologies, is usually used to detect fractures.

1.3 Fracture Models and Assumptions

The fracture model is a link between subsurface fracture characteristics and surface seismic data (Sava 2004). There are three popular fracture models that are widely investigated: the penny-shaped crack model (Hudson 1980, 1981), the linear slip model (Schoenberg 1980, 1983; Schoenberg and Muir 1989) and a combination of the penny-shaped crack model and the linear slip model (Hudson 1997). The first model describes fractures as ellipsoidal cracks, with the parameters being the crack density and the aspect ratio of the oblate spheroidal cracks. The second model specifies fractures as infinitely extended weakness planes, with the parameters being the normal and tangential fracture compliances. These two models form the basis of an effective medium theory and are elastically equivalent to each other (Schoenberg and Douma 1988; Li 1998), even though the models have different descriptions for the fractures. The third model depicts the fracture as an infinite plane with two rough surfaces in partial contact in which the noncontact parts are represented as a planar distribution of penny-shaped cracks, while elsewhere the parts in contact support the ambient pressure and friction. Because the fracture opening is very small with respect to the seismic wavelength, the geometric details of the fracture such as shape and microstructure are often neglected in a fracture detection study. Hence, the models related to the penny-shape model are outside of the scope of this book. The linear slip model will be studied in detail and used as the fracture model in forward modeling and backward inversion. In addition, Pyrak-Nolte et al. (1990) and Hsu and Schoenberg (1993) confirmed the validity of the linear slip model by some laboratory experimental verifications.

The effective medium theory of Backus (1962) is developed from averaging the properties of multiple isotropic thin layers to directly make up a single composite anisotropic medium. The description of the composite medium is based on the relative fractions and properties of the thin layers. In 1980, with the effective medium theory, Michael Schoenberg described a physical mechanism of the linear slip interface: for an isotropic thin layer embedded in a homogeneous isotropic host medium, the thin layer can be treated as a linear slip interface once the layer thickness is much thinner than wavelength and its impedance is very small compared to the host medium. To some degree, this linear slip interface is equivalent to the fracture interface.

Schoenberg and Muir (1989) extended Backus's averaging approach to develop a matrix formalism for calculating the properties of a fractured medium that is made up of fracture and host medium. The composite medium, with five independent moduli, exhibits the properties of a transversely isotropic medium with a symmetric axis (Helbig 1999, 2009; Nichols 1989; Hood 1989, 1991) that may be different from that of the initial host medium. According to Schoenberg and Muir (1989), the fractured medium can be simplified by performing a composition and a decomposition calculation as follows:

$$\text{Fracture} + \text{Host medium} \leftrightarrow \text{Fractured medium}$$

Therefore, a horizontally fractured medium is composed of a horizontal fracture and a homogeneous isotropic host medium and exhibits the properties of a transversely isotropic medium with a vertical symmetric axis (a VTI medium). A vertically fractured medium is formed by a vertical fracture and a homogeneous isotropic host medium and exhibits the properties of a transversely isotropic medium with a horizontal symmetric axis (a HTI medium). The simplest orthogonally fractured medium is assembled by a horizontal fracture and a HTI host medium, or a vertical fracture and a VTI host medium, or two orthorhombic fractures embedded into an isotropic host medium.

Schoenberg's linear slip model simulates the fracture interface satisfying the nonwelded contact boundary conditions (Schoenberg 1980; Krebes 1987) that constrain all seismic waves at the fracture interface. The nonwelded contact boundary conditions state that the dynamic stresses of the wave quantities are continuous across the fracture boundary, whereas the kinematic displacements of the wave quantities are discontinuous across the fracture boundary. Thus the reflection and transmission coefficients of waves at the fracture interfaces (Chaisri and Krebes 2000) have expressions different from those at welded boundaries (Aki and Richards 1980).

1.4 Forward Modeling of the Fracture

Seismic forward modeling is a key connection between the model and seismic response, and it can predict results, and enhance interpretation and inversion (Lines and Newrick 2004). Finite-difference (FD) methods are widely used in seismic forward modeling to numerically simulate elastic wave propagation in the medium (Kelly et al. 1976; Virieux 1986; Lines et al. 1999; Zhang 2005). Kelly et al. (1976) presented a so-called homogeneous approach for finite-difference formulations of the elastic wave equation. In this approach, boundary conditions at interfaces are imposed explicitly (Slawinski and Krebes 2002a, b). With the linear slip interface model, Coates and Schoenberg (1995) introduced an equivalent medium approach to treat fractures. In their approach, the grid containing a fracture is replaced by the grid with equivalent anisotropic properties. The forward modeling approach of Slawinski and Krebes (2002a, b) employed boundary conditions explicitly, and additional fictitious nodes were incorporated by them. Their approach adapted the homogeneous formulation of Kelly et al. (1976) by treating the fracture as a non-welded contact interface satisfying the linear slip displacement discontinuity conditions. The fictitious nodes have the same physical grid locations as the real grid nodes, but it is convenient to explicitly apply the boundary conditions. In 2009, Zhang and Gao presented a numerical modeling study for elastic wave propagation in a 3-D medium with 2-D fractures by imposing the boundary conditions explicitly. Cui (2013a) modeled a horizontal fracture, a vertical fracture and an orthorhombic fracture by using the homogeneous formulation approach with fictitious nodes and explicit boundary conditions (Slawinski and Krebes 2002a, b). In addition, in a CHOPS (Cold Heavy Oil Production with Sand) study, a fractal pattern of the wormholes in the medium is modeled by applying 3D FD schemes with the same modeling approach (Cui 2013b).

Employing the homogeneous formulation approach with additional fictitious nodes and explicit boundary conditions, theoretically, finite-difference schemes using nonwelded contact boundary conditions are presented individually in this book for the horizontally fractured medium, the vertically fractured medium, and the orthogonally fractured medium. These new finite-difference schemes express the wave propagation in the fractured medium, in which the normal equation of wave motion governs the wave propagation in the host medium and the nonwelded contact boundary conditions constrain the waves at the fracture interface.

The new finite-difference schemes that generate the synthetic seismograms for the different fractured media are developed using MATLAB. These modeled seismograms indicate that the fractures are detectable because fractures can act as reflectors and can therefore significantly affect seismic wave propagation. The amplitude levels of the events in multicomponent seismic data reveal the direction of the fracture. Therefore, knowledge of how fractures are represented in seismic data can enhance fracture detection.

1.5 AVO Inversion of a Fractured Medium

Using any possible means to infer the lithology and fluid properties in a reservoir is an ultimate goal of oil and gas exploration. AVO (amplitude variation with offset) inversion attempts to use the amplitudes of available surface seismic data to estimate the reflectivity of the density, the P-wave velocity and the S-wave-velocity of the earth-model (Lines and Newrick 2004).

Since the 1960s geophysicists have discovered that gas deposits are related to amplitude anomalies on stacked sections known as "bright spots", and many geoscientists have been aware that the surface-recorded seismic amplitudes can be associated with rock properties by studying the Zoeppritz equations (1919). For example, Koefoed (1955), via a study of the exact Zoeppritz equations, pointed out the relationship of AVO to Poisson's ratio across a boundary. Bortfeld (1961) simplified Zoeppritz's equations and showed how the reflection amplitudes depend in a simpler way on an incident angle and the physical parameters of the medium. In 1980, Aki and Richards presented the AVO equations in terms of the P-velocity, S-velocity and density reflectivity of the medium. Ostrander (1984), and Fatti et al. (1994) illustrated the interpretational benefits of AVO in predicting and mapping hydrocarbons. Shuey (1985) specified the AVO equations in terms of the zero-offset of the P-wave reflectivity and a so-called AVO gradient. Smith and Gidlow (1987) rearranged the Aki and Richards's equations and applied an empirical relationship (Gardner et al. 1974) to approximate the AVO equationsand express them in terms of the P-wave and S-wave velocity reflectivity of the medium. Rutherford and Williams (1989) proposed a classification scheme for the AVO anomalies, with further modifications made by Rossand Kinman (1995) and Castagna and Swan (1997). Goodway et al. (1997) used the AVO method to indicate lithology and map porosity. In 2002, Rueger used Shuey's approach (Shuey 1985) to unravel an AVAZ equation and showed that the reflection coefficients are directly impacted by both incident angles and azimuths. Downton (2011) contributed to the problem of the AVAZ equation's relation to fracture weakness parameters through a method of azimuthal Fourier Coefficients. However, these approximate AVO equations are entirely based on an assumption of perfect welded contact regardless of the fractured medium.

Schoenberg (1980), Pyrak-Nolte (1990a, 1990) and Chaisri and Krebes (2000) derived the exact formulae for the reflection and transmission coefficients with nonwelded contact boundary conditions that point to fracture issues. Based on these equations, Cui et al. (2013c) derived approximate AVO equations and implemented AVO inversion to estimate all rock properties related to the fracture from seismic amplitudes.

Because AVO inversion is based on the amplitude variation over a range of the offsets from the surface seismic data, the seismic data must accurately preserve the true amplitudes in correspondence to geological factors, rather than containing any signs of non-geological bodies and artifacts from the acquisition. Thus some pre-conditioning processing, such as deconvolution, noise attenuation, 5D interpolation, and shear-wave layer stripping should be applied to data to keep the true amplitude variation.

New equations for reflection and transmission coefficients which include the tangential and normal fracture compliances are presented here, and are based on nonwelded contact boundary conditions. The equations have a pattern similar to that of the original Zoeppritz equations. The equations reduce to the original Zoeppritz equations with the assumption of welded contact boundary conditions (when the tangential and normal fracture compliances vanish), or the fractured medium equations when there is no impedance contrast in the host medium. Thus, the new equations allow for the descriptions of the subsurface geological bodies with both the welded and nonwelded contact boundary conditions discussed above.

New approximate equations are presented for the AVO inversion of a horizontally fractured medium. As we know, a fractured medium can be decomposed into a fracture and a host medium, vice versa. The new approximate AVO equations consist of the elastic reflectivity of the welded contact part caused by the impedance contrast interfaces and the reflectivity of the nonwelded contact part produced by the fracture. Thus, an accurate inversion of the elastic reflectivity of the host medium of the fractured medium with impedance contrast should employ the new AVO equations because they provide a way to separate the fracture effects from the seismic data. In other words, input data should not include the contamination from the fracture when we attempt to invert elastic properties for the host medium. Also, the fracture properties of the fractured medium, which is a fracture embedded in a uniform isotropic host medium, can be estimated by using the new AVO inversion equations, whereas conventional AVO inversion cannot achieve these results because the conventional AVO equations are just for the reflectivity inversion of a welded contact medium. Therefore, for the inversion of any medium properties, one should apply the new AVO inversion equations. Especially, for the inversion of all the properties related to the fractures, one should apply the new AVO inversion equations for correct reservoir delineation.

1.6 Outline of This Book

The structure of this book follows closely the description of the methodology introduced in the previous sections. The first part of this book in Chap. 2 focuses on the rock physics theory about fractures, including the fracture mechanism, the composition of the fractured medium, and the boundary conditions of the fracture. The second part in Chap. 3 focuses on the seismic modeling of the fracture to illustrate that the fractures are detectable and fracture responses can be identified in the seismic data. The last part of this book in Chap. 4 presents new AVO equations for correctly estimating all the rock properties of a fractured and unfractured medium of the reservoir.

Chapter 2 of this book describes some indispensable fundamental concepts related to fractures. The concepts of stress and strain from rock physics theories are reviewed to clarify the stages of rock deformation from reversible elasticity to irreversible permanent fractures. The fracture parameters such as height, length, opening, and orientation are used to describe existing fractures. In geoscience, the fracture orientation is a more important parameter than the fracture opening, and is parallel to the direction of maximum compressive stress and normal to the direction of minimum compressive stress, and fracture formation is different with the formation of the fault. Fracture detection usually employs an integrated method, involving geology, petrophysics, and geophysics, while geophysical detection is a big challenge that demands developing a method for enhancement of fracture detection and correct reservoir delineation, even though the petrophysical data are the most reliable. Compared to the intrinsic anisotropy medium, the fractured medium exhibits an induced anisotropic feature that is consistent with previous theoretical work and field observations. Therefore, different symmetric planes of an anisotropic medium have different anisotropic symbols, and also different anisotropic parameter symbols correspond to fracture-induced anisotropy in terms of tangential and normal fracture compliances for different fractured media. Through the Schoenberg and Muir calculus study, the fractured medium can be composed of fractures embedded in a host medium, and the composite medium, with five independent moduli, possesses anisotropy characteristics. With the effective medium assumption, a linear slip model is chosen to simulate fractures, because the fracture is an extended weakness interface regardless of the fracture shape and microstructure, and all seismic waves at the fracture interface satisfy the nonwelded contact boundary conditions which state that the dynamic stresses are continuous across the fracture, but the kinematic displacements are discontinuous across the fracture.

A table for the types of media contact and a table for boundary conditions are given in this chapter. For the table of the media contact, a “fractured medium” means that the host medium has no impedance contrast, whereas “fractured medium with impedance contrast” indicates that the host media are homogenous and that there is an impedance contrast, and the fracture interface and the impedance interface are superposition. In the boundary conditions table, the terms z-normal boundary, x-normal boundary, and y-normal boundary refer to a horizontal boundary, a vertical boundary that is normal to the x-axis and a vertical boundary that is normal to the y-axis, respectively.

Chapter 3 focuses on the seismic forward modeling of the fracture in order to study fracture representations in seismic data. In this chapter, the finite-difference forward modeling formulation of the homogeneous approach has been studied and selected for fracture modeling. Meanwhile, fictitious grids are introduced into the FD scheme for the purpose of explicitly imposing boundary conditions at the interface. Based on the finite-difference scheme for fracture modeling, we individually derived new formulae of FD modeling for the horizontally fractured medium, the vertically fractured medium and the orthogonally fractured medium. Especially, for a FD grid, the scheme specifies only one nonwelded contact boundary for modeling the fracture interface, and the rest of the three boundaries are welded contact boundaries. This is different from the FD formulae of Slawinski (1999) in which all boundaries are allowed to be nonwelded contact for modeling the fracture interface in a FD grid. In the last part of this chapter, under a fractured medium stability condition, we implement a MATLAB script for running the new fracture FD schemes with a Ricker source wavelet to generate seismograms from the horizontally fractured medium, the vertically fractured medium and the orthogonally fractured medium. The results indicate that the fractures are predictable and visible, even if the fractured medium has no impedance contrast. The synthetic seismic data for the fractured medium demonstrate the characteristics of the fracture-induced anisotropy, in terms of traveltime and amplitude, providing us with knowledge of the fractures which can be used to enhance fracture detection.

Chapter 4 emphasizes the derivation of equations for the PP and PS reflection and transmission coefficients, the approximate PP AVO (Amplitude Variation with Offset) equations for the horizontally fractured medium, and conducts inversion to estimate all rock properties related to horizontal fractures. In this chapter, first, we reviewed some published AVO equations based on the assumption of a welded contact medium, and some seismic data preconditioning techniques such as deconvolution, noise attenuation, 5D interpolation, and shear-wave layer stripping to preserve the true amplitudes corresponding to geological factors. Second, with the inclusion of the parameters of the normal and tangential fracture compliances, we present the reflection and transmission coefficient expressions for PP and PS waves at the horizontal fracture interfaces that satisfy the nonwelded contact boundary conditions, and an exact solution for the PP reflection coefficients for the horizontally fractured medium. Third, we derive new approximate AVO equations that can be divided into a welded contact part and a nonwelded contact part. The welded contact part agrees with the Aki and Richard's AVO equations. The nonwelded contact part, derived in a similar way, using the same conditions for approximation, results in eight items related to the fracture of which two items are independent of the host media. The new AVO equations reduce to the conventional AVO equations to predict the reflectivity of the rock when the tangential and normal compliances of the fracture parameters vanish, or if the equations are only for the inversion of the fracture properties when the medium has no impedance contrast. In other words, the new AVO equations are able to correctly and individually estimate all rock properties, such as the fracture properties, and host media properties and fractured medium with impedance contrast properties, whereas with the conventional AVO equation, it is hard to estimate all rock properties related to fractures. Finally, we apply a generalized linear inversion algorithm (GLI) to invert for all rock properties related to the fractured media, to demonstrate the difference that results when the conventional AVO equation is used to invert for the properties for the same fractured media, and to invert for the tangential and normal compliances (fracture parameters) from seismic data as well.

1.7 Conclusions

The key concepts of the related fracture and the fractured media are studied to meet the imminent industrial demands as described above, i.e., to deeply understand fracture representations in seismic data in order to enhance the geophysical prediction of the fracture distribution, and to accurately estimate all the medium properties related to fractures for the correct delineation of the reservoir. This ultimately benefits reservoir and mining engineers, geologist and geophysicist with respect to optimizing the hydrocarbon reservoir, well performance, and safety mining.

References

Aguilera, R. (1998). Geologic aspects of naturally fractured reservoirs. *The Leading Edge, 17,* 1667–1670.

Aki, K., & Richards, P. G. (1980). *Quantitative seismology: theory and methods* (Vol. 1). W.H: Freeman and Company.

Anderson, E. M. (1905). The dynamics of faulting. *Transactions of the Edinburgh Geological Society, 8,* 387–402.

Anderson, E. M. (1942/1951). The dynamics of faulting and dyke formation with application to Britain. Oliver & Boyd: Edinburgh.

Backus, G. E. (1962). Long-wave elastic anisotropy produced by horizontal layering. *Journal of Geophysical Research, 67,* 4427–4440.

Bakulin, A., Grechka, V., & Tsvankin, I. (2000a). Estimation of fracture parameters from reflection seismic data-Part I: HTI model due to a single fracture set. *Geophysics, 65*(6), 1788–1802.

Bakulin, A., Grechka, V., & Tsvankin, I. (2000b). Estimation of fracture parameters from reflection seismic data-Part II: fractured models with orthorhombic symmetry. *Geophysics, 65*(6), 1803–1817.

Bakulin, A., Grechka, V., & Tsvankin, I. (2000c). Estimation of fracture parameters from reflection seismic data-Part III: fractured models with monoclinic symmetry. *Geophysics, 65*(6), 1818–1830.

Bortfeld, R. (1961). Approximation to the reflection and transmission coefficients of plane longitudinal and transverse waves. *Geophysical Prospecting, 9,* 485–503.

Cary, P. (2008). The impact of shear-wave splitting on the imaging of P-S data: CSPG CSEG CWLS convention.

Castagna, J. P., & Swan, H. W. (1997). Principles of AVO crossplotting. *The Leading Edge, 16*(04), 337–342.

Chaisri, S., & Krebes, E. S. (2000). Exact and approximate formulas for P-SV reflection and transmission coefficients for a nonwelded contact interface. *Journal of Geophysical Research, 105,* 28045–28054.

Coates, R. T., & Schoenberg, M. (1995). Finite-difference modeling of faults and fractures, *Geophysics*, *60*, 1514–1526.

Crampin, S. (1985). Evidence for aligned cracks in the Earth's crust. *First Break, 3*(3), 12–15.

Cui, X., Lines, L, R., & Krebes, E. S. (2013a). Numerical modeling for different types of fractures, *GeoConvention:* Integration.

Cui, X., Lines, L. R., & Krebes, E. S. (2013b). Numerical modeling fractures and fractal network feature-wormhole of CHOPS. SEG Houston 2013 annual meeting. DOI http://dx.doi.org/10.1190/segam2013-1179.1

Cui, X., Lines, L. R., & Krebes, E. S. (2013c). Seismic modeling and inversion for refection from fractures. CREWES Research Report (2013).

Downton, J., Roure, B., & Hunt, L. (2011). Azimuthal fourier coefficients: CSEG recorder, *36*(10), 22–36.

Dunphy, R., & Campagna, D. J. (2011). Fractures, Elastic Moduli & Stress: geological controls on hydraulic fracture geometry in the horn river basin. In *Conference Paper at Energy, Environment, Economy Recovery* SCPG/CSEG/CWLS convention in Calgary.

Fatti, J. L., Smith, G. C., Vail, P. J., Strauss, P. J., & Levitt, P. R. (1994). Detection of gas in sandstone reservoirs using AVO analysis: A 3-D seismic case history using the Geostack technique. *Geophysics, 59,* 1362–1376.

Gardner, G. H. F., Gardner, L. W., & Gregory, A. R. (1974). Formation velocity and density—The diagnostic basics for stratigraphic traps. *Geophysics, 39,* 770–780.

Gary, D. (2008). Fracture detection using 3D seismic azimuthal AVO, *CSEG Recorder.*

Goodway, B., Chen, T., & Downton, J. (1997). Improved AVO fluid detection and lithology discrimination using lame petrophysical parameters; "λρ", "μρ" and "λμ fluid stack", from P and S inversions. CSEG expanded abstracts, 148–151.

Helbig, K. (1999). Layer-induced elastic anisotropy—Part 1: forward relations between constituent parameters and compound medium parameters. *Revista Brasileira de Geofsica*, *16*, 103–112.

Helbig, K., & Carcione, J. M. (2009). Anomalous polarization in anisotropic media. *European Journal of Mechanics–A/Solids, 28,* 704–711.

Hood, J. (1991). A simple method for decomposing fracture-induced anisotropy. *Geophysics, 56,* 1275–1279.

Hood, J. A. & Schoenberg, M. (1989). Estimation of vertical fracturing from measured elastic moduli, *Journal of Geophysical Research*, *94*(15), 611–618.

Hsu, C.-J. & Schoenberg, M. (1993). Elastic waves through a simulated fracture medium, *Geophysics*, *58*, 964–977.

Hudson, J. A. (1980). Overall properties of a cracked solid. *Mathematical Proceedings of the Cambridge Philosophical Society.88*, 371–384.

Hudson, J. A. (1981). Wave speeds and attenuation of elastic waves in material containing cracks. *Geophysical Journal of the Royal Astronomical Society, 64,* 133–150.

Hudson, J. A., & Harrison, J. P. (1997). Engineering rock mechanics: an introduction to the principles (p. 456).New York, Pergamon.

Jaeger, J. C. (1969). Elasticity, fracture and flow, with engineering and geological applications (3th ed., p. 268). London: Chapman and Hall.

Kelly, K. R., Ward, R. W., Treitel, S.,. & Alford, R. M. (1976). Synthetic seismograms finite-difference approach, *Geophysics*, *41*(1), 2–27.

Koefoed, O. (1955). On the effect of Poisson's ratios of rock strata on the reflection coefficients of plane waves. *Geophysical Prospecting, 3,* 381–387.

Krebes, E. S. (1987). Reflection and transmission at plane boundaries in nonwelded contact, *Journal of the Canadian Society of Exploration Geophysicists*, *23*, 66–27.

Lawton, D. C. (1999). Slip-slidin' away—Some practical implications of seismic velocity anisotropy on depth imaging. In L.R. Lines, D.C. Lawton & S.H. Gray (Eds.), Depth imaging of foothills seismic data, *Canadian Society of Exploration Geophysicists.*

Lines, L. R., Chen, S., Daley, P. F., & Embleton, J. (2003). *Seismic pursuit of wormholes: The Leading Edge, 22,* 459–461.

Lines, L. R., & Daley, P. F. (2007). Seismic detection of cold production footprints in heavy oil extraction. *Journal of Seismic Exploration, 15*(4), 333–344.
Lines, L. R., Daley, P. F., Bording, R. P., & Embleton, J. E. (2008). The resolution and detection of "subseismic" high-permeabilityzones in petroleum reservoirs, *The Leading Edge, 27*(5), 664–669.
Lines, L. R., & Newrick, R. T. (2004). *Fundamentals of geophysical interpretation: Society of Exploration Geophysicists publication*. Oklahoma: Tulsa.
Lines, L. R., Slawinski, R., & Bording, R. P. (1999). Short note—A recipe for stability of finite-difference wave-equation computations, *Geophysics*, *64*(3), 967–969.
Massonnat, G. J., Umbhauer, F., & Odonne, F. (1994). The use of 3-D seismic in the understanding and monitoring of waterflooding in a naturally fractured gas reservoir. *AAPG Bulletin, 78*(7), 1155.
Nelson, R. A. (1985). Geologic analysis of naturally fractured reservoirs. Houston: Gulf Pub. Co., Book Division, c1985.
Nichols, D., Muir, F., & Schoenberg, M. (1989). Elastic properties of rocks with multiple fracture sets. In *59th Annals of International. Mtg. Expanded abstracts* (pp. 471–474). Society of Exporation Geophysical.
Ostrander, W. J. (1984). Plane-wave reflection coefficients for gas sands at nonnormal angles of incidence. *Geophysics, 49,* 1637–1648.
Price, N. J. (1966). Fault and joint development in brittle and semi-brittle rock(p. 176). Oxford: Pergamon Press.
Priest, S. D., & Hudson, J. A. (1976). Discontinuity spacings in rock. *International Journal of Rock Mechanics Mining Sciences and Geomechanics Abstracts, 13*, 135–148.
Pyrak-Nolte, L. J., Myer, L.R., & Cook, N.G.W. (1990a). Transmission of seismic waves across single natural fractures. *Journal of Geophysical Research, 95*, 8617–8638.
Pyrak-Nolte, L. J., Myer, L. R., & Cook, N. G. W. (1990b). Anisotropy in seismic velocities and amplitudes from multiple parallel fractures. *Journal Geophysical Research, 95*(B7), 11345–11358.
Richard, B., Poplavskii, K., Wright, C. (2009). Compensation for shear-wave anisotropy effects above a Heavy Oil Reservoir: CSPG CSEG CWLS convention.
Rider, M. (2002). *The geological interpretation of well logs* (2nd ed.). Sutherland: Rider-French Consulting Ltd.
Romm, E. S., (1985). Structural models of rocks pore space. Nedra (in Russian).
Ross, C. P., & Kinman, D. L. (1995). Nonbright-spot AVO; two examples. *Geophysics, 60,* 1398–1408.
Rueger, A. (2002). *Reflection coefficients and azimuthal AVO Analysis in anisotropic media: SEG geophysical monograph series number10: Soc.* Geophys: Expl.
Rutherford, S. R., & Williams, R. H. (1989). Amplitude-versus-offset variations ingas sands, *Geophysics*, *54*, 680–688.
Sava, D. C. (2004). *Quantitative data integration for fracture characterization using statistical rock physics, Ph. D. thesis*, Stanford University.
Schlumberger. (1989). Log interpretation principles/applications. Schlumberger Educational Services.
Schoenberg, M. (1980). Elastic wave behavior across linear slip interfaces. *Journal of the Acoustical Society of America, 68,* 1516–1521.
Schoenberg, M. (1983). Reflection of elastic waves from periodically stratified media with interfacial slip. *Geophysical Prospects, 31,* 265–292.
Schoenberg, M., & Douma, J. (1988). Elastic wave propagation in media with parallel fractures and aligned cracks, *Geophysical Prospects*, *36*, 571–590.
Schoenberg, M., & Muir, F. (1989). A calculus for finely layered anisotropic media. *Geophysics, 54,* 581–589.
Schultz, R. A., & Fossen, H. (2008). Terminology for structural discontinuities. *American Association of Petroleum Geologists Bulletin, 92,* 853–867.
Shuey, R. T. (1985). A simplification of the Zoeppritz equations. *Geophysics, 50,* 609–614.

Singhal, B. B. S., & Gupta, R. P. G. (1999). Applied hydrogeology of fractured rocks. Kluwer Academic Publishers, Dordrecht, The Netherlands, p. 400.

Slawinski, R. A. (1999). *Finite-difference modeling of seismic wave propagation in fractured media: ph.d. thesis*, The University of Calgary.

Slawinski, R. A., & Krebes, E. S. (2002a). Finite-difference modeling of SH-wave propagation in nonwelded contact media. *Geophysics, 67,* 1656–1663.

Slawinski, R. A., & Krebes, E. S. (2002b). The homogeneous finite difference formulation of the P-SV wave equation of motion: Studia Geophysica et Geodaetica (Impact Factor: 0.75). 01/2002; 46(4), pp. 731–751. DOI: 10.1023/A:1021133606779.

Smith, G. C., & Gidlow, P. M. (1987). Weighted stacking for rock property estimation and detection of gas. *Geophysical Prospects, 35,* 993–1014.

Teng, L. (1998), Ph.D. Dissertation, Stanford University.

Thomsen, L. (1995). Elastic anisotropy due to aligned cracks in porous rock, *Geophysical Prospecting*, *43*(6), 805–829.

Virieux, J. (1986). P-SV wave propagation in heterogeneous media: Velocity-stress finite-difference method, *Geophysics*, *51*(4), 889–901.

Yu, T. R., & Telford, W. M. (1973). An ultrasonic system for fracture detection in rock faces. *Canadian Mineral Metal Bull., 66*(729), 96–101.

Yuan, J. Y., Tremblay, B., & Babchin, A., (1999). A wormhole network model of cold production in heavy oil: SPE Paper 54097.

Zhang, J. (2005). Elastic wave modeling in fractured media with an explicit approach. *Geophysics, 70,* T75–T85.

Zhang, J., & Gao, H. (2009). Elastic wave modelling in 3-D fractured media: An explicit approach. *Geophysical Journal International, 177,* 1233–1241.

Zoeppritz, K. (1919). Erdbebenwellen VIIIB, On the reflection and propagation of seismic waves: Göttinger Nachrichten, I, 66.

Chapter 2
Geological Fractures and Geophysical Assumptions

Abstract Studying geological fractures from seismic data not only requires performing forward modeling and backward inversion, but also required an understanding of the rock mechanisms and some geophysical hypotheses related to fractures. In this chapter, many indispensable fundamental concepts are reviewed that are associated with fracture formations, fracture types, fracture parameters, fracture detections, fracture equivalent hypotheses (effective fractures), fractured media composition, and decomposition and fracture interface boundary conditions especially, rock deformations, Anderson's principal stresses classification (Trans Edinb Geol Soc 8: 387–402, 1905), petrophysics technologies, linear slip theory (Schoenberg in J Acoust Soc Am 68: 1516–1521, 1980), Schoenberg and Muir calculus theory (Geophysics 54: 581–589, 1989), welded contact boundary conditions, and nonwelded contact boundary conditions are explored and integrated to form a substantial body of material supporting the geological fractured media forward modeling in Chap. 3 and seismic inversion in Chap. 4 of this book.

Keywords Stress–strain tensors · Rock deformation · Stiffness-compliances · Fracture formation · Fracture parameters · Fracture detection · Well logs · Backus average theory · Linear slip theory · Schoenberg-Muir theory · Boundary conditions · Host medium · Fractured medium · Isotropy · Intrinsic anisotropy · Induced anisotropy

2.1 Rock Deformations

2.1.1 Stress Tensor

Stress and strain tensors are keys for deeply understanding the fracture mechanisms. In earth science, permanently fractured rock usually experiences three successive stages when it is continually subjected to forces.

Traction is the force per area acting within a deformable material.

© Springer Nature Singapore Pte Ltd. 2018

X. Cui et al., *Seismic Forward Modeling of Fractures and Fractured Medium Inversion*, Springer Geophysics, DOI 10.1007/978-981-10-3584-5_2

$$\boldsymbol{T} = \frac{\boldsymbol{F}}{S}, \tag{2.1a}$$

$$T_i = \sum_{j=1}^{3} \sigma_{ij} n_j, \tag{2.1b}$$

where $\boldsymbol{T}$ is traction as the force per unit surface area, $\boldsymbol{F}$ is a force, and S is an area subjected to the force, $\boldsymbol{\sigma}$ is stress tensor, n is surface. $i, j = 1, 2, 3$. In general, an arbitrary stress $\boldsymbol{\sigma}$ is a second rank tensor consisting of nine components, and can be expressed in matrix form as

$$\boldsymbol{\sigma} = \begin{bmatrix} \sigma_{11} & \sigma_{12} & \sigma_{13} \\ \sigma_{21} & \sigma_{22} & \sigma_{23} \\ \sigma_{31} & \sigma_{32} & \sigma_{33} \end{bmatrix}. \tag{2.2}$$

σ_{ij} are the components of the tensor in the Cartesian coordinate system. Index "*i*" describes the direction of the force component, and index "*j*" denotes the direction that is perpendicular to the surface on which the force acts (Krebes 2006). For example, the stress component σ_{12} describes a force component parallel to (1)-axis and acting on the surface (the (1) & (3)-plane) that is normal to (2)-axis. In general, σ_{11}, σ_{22} and σ_{33} are so-called normal stresses because these components are normally acting to the surfaces, while $\sigma_{12}, \sigma_{13}, \sigma_{21}, \sigma_{23}, \sigma_{31}$ and σ_{32} are so-called shear stresses that are tangentially acting on the surfaces. The components of the stress are shown in Fig. 2.1 on which bold arrows indicate the normal stress components and light arrows indicate the tangential ones. The stress tensor $\boldsymbol{\sigma}$ is commonly expressed in terms of Voigt notation $(11 \rightarrow 1, 22 \rightarrow 2, 33 \rightarrow 3, 23 \rightarrow 4, 31 \rightarrow 5, 12 \rightarrow 6)$ which also makes use of the symmetry of the tensor $(13 \rightarrow 31, 23 \rightarrow 32, 12 \rightarrow 21)$ in the Cartesian coordinate system.

$$\begin{aligned} \boldsymbol{\sigma} &= \begin{bmatrix} \sigma_{11} & \sigma_{22} & \sigma_{33} & \sigma_{23} & \sigma_{31} & \sigma_{12} \end{bmatrix}^{\mathrm{T}} \\ &= \begin{bmatrix} \sigma_1 & \sigma_2 & \sigma_3 & \sigma_4 & \sigma_5 & \sigma_6 \end{bmatrix}^{T} \end{aligned} \tag{2.3}$$

2.1.2 Strain Tensor

The components of the strain tensor, $\boldsymbol{\varepsilon}$, are defined in terms of the relative changes in the displacement components of the deformed material subjected to the stresses. Similar to the stress tensor, the strain tensor is symmetric, and has nine components: the normal strains $\varepsilon_{11}, \varepsilon_{22}, \varepsilon_{33}$ and the shear strains $\varepsilon_{12}, \varepsilon_{13}, \varepsilon_{21}, \varepsilon_{23}, \varepsilon_{31}, \varepsilon_{32}$. The normal strains measure relative changes in displacement along a specific direction. For instance, ε_{22} can be given as

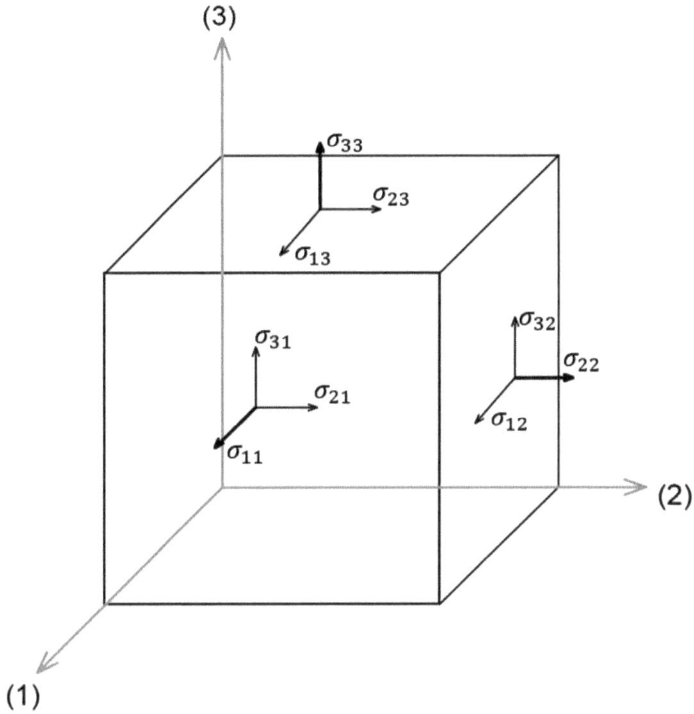

Fig. 2.1 A sketch of the nine components of the stress tensor at a point in a Cartesian coordinate system. The *bold arrows* indicate the normal components of the stress. The *light arrows* indicate the tangential components of the stress

$$\varepsilon_{22} = \frac{\Delta u_2}{\Delta l_2}, \tag{2.4}$$

where, Δl_2 is the original length and Δu_2 is the change of displacement along (2)-axis direction (Fig. 2.2a).

The shear strains simply measure the material distortion of the shape in (small) angles with respect to a specific direction. Figure 2.2b shows simultaneous shear strains along both the (2)-axis and the (3)-axis. The change in shape of the material is due to the applied force that can be described by

$$\varepsilon_{32} = \frac{1}{2}(\phi_2 + \phi_3) \approx \frac{1}{2}\left(\frac{\Delta u_3}{\Delta l_2} + \frac{\Delta u_2}{\Delta l_3}\right), \tag{2.5}$$

where the definitions of Δl_2 and Δu_2 are the same as in Eq. (2.4). Δl_3 is the original length and Δu_3 is the change in displacement along (3)-axis direction.ϕ_2 and ϕ_3 are material shape distortions in angles related to (2)-axis and (3)-axis, respectively (Fig. 2.2b).

Fig. 2.2 **a** A sketch of the normal strain showing a size change in the Cartesian coordinate system. $\Delta \boldsymbol{l}_2$ is the original length and $\Delta \boldsymbol{u}_2$ is the change in displacement along the 2-axis direction.
b Diagram of the tangential (shear) strains in shape change in the Cartesian coordinate system. $\Delta \boldsymbol{l}_2$, $\Delta \boldsymbol{l}_3$ are original lengths and $\Delta \boldsymbol{u}_2$, $\Delta \boldsymbol{u}_3$ are changes in displacement along (2)-axis, (3)-axis direction respectively. $\boldsymbol{\phi}_2$ and $\boldsymbol{\phi}_3$ are material shape distortions in angles respect to (2)-axis and (3)-axis, respectively

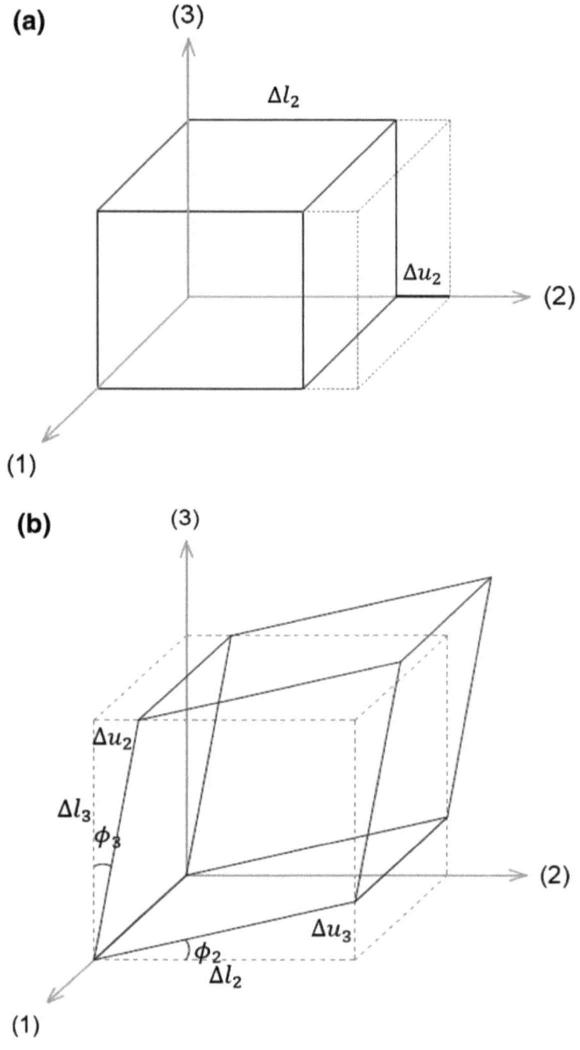

The strain tensor $\boldsymbol{\varepsilon}$, can also be expressed in Voigt notation rule which makes use of the symmetry in the Cartesian coordinate system as well, i.e.,

$$\begin{aligned}\boldsymbol{\varepsilon} &= \begin{bmatrix} \varepsilon_{11} & \varepsilon_{22} & \varepsilon_{33} & \varepsilon_{23} & \varepsilon_{31} & \varepsilon_{12} \end{bmatrix}^{\mathrm{T}} \\ &= \begin{bmatrix} \varepsilon_{1} & \varepsilon_{2} & \varepsilon_{3} & \varepsilon_{4} & \varepsilon_{5} & \varepsilon_{6} \end{bmatrix}^{T}. \end{aligned} \tag{2.6}$$

In the infinitesimal limit, the strain is given by (Krebes 2006)

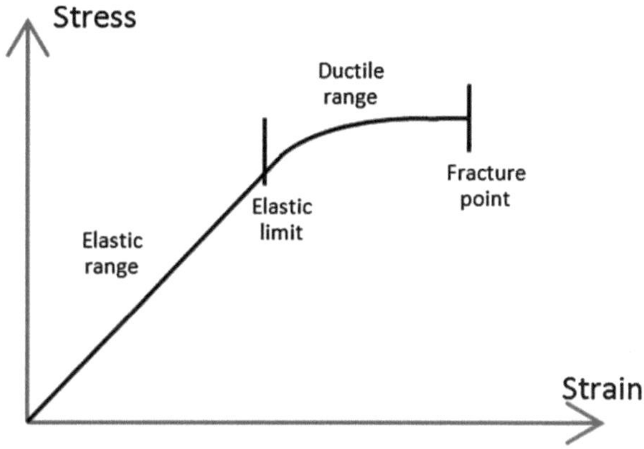

Fig. 2.3 Diagram of stress–strain curve and rock deformation stages. The stages involve reversible elastic deformation and irreversible ductile deformation and fracture deformation

$$\varepsilon_{ij} = \frac{1}{2}(\phi_j + \phi_i) \approx \frac{1}{2}\left(\frac{\partial u_i}{\partial x_j} + \frac{\partial u_j}{\partial x_i}\right). \tag{2.7}$$

The rock strains represent either reversible deformations where an object can return to its original size and shape once the exerting stress is removed or irreversible deformations where an object is permanently distorted even if the exerting stress disappears. The deformation stages actually depend on the ratio of the stress and strain. Figure 2.3 shows a stress and strain function and the deformations (Nelson 2003).

2.1.3 Stages of the Rock Deformation

For a small stress exerted on the rock, Hooke's law entirely describes the linear elastic behavior on the first elastic stage of the reversible rock deformation (Fig. 2.3). The general mathematical equation for Hooke's law is

$$\sigma_{ij} = \sum_{k=1}^{3}\sum_{l=1}^{3} C_{ijkl}\varepsilon_{kl}, \tag{2.8}$$

$$C_{ijkl} = \lambda\delta_{ij}\delta_{kl} + \mu(\delta_{il}\delta_{jk} + \delta_{ik}\delta_{jl}), \tag{2.9}$$

where the quantity C_{ijkl} is called the material stiffness tensor or material modulus tensor that is a measurement of the resistance offered by an elastic body to deformation, and where $i, j, k, l = 1, 2, 3$. λ and μ are known as the Lame physical parameters, and δ_{ij} is the Kronecker delta:

$$\delta_{ij} = \begin{cases} 1, & i = j \\ 0, & i \neq j \end{cases}. \tag{2.10}$$

With Voigt notation, Hooke's law in Eq. (2.8) can be rewritten as

$$\begin{bmatrix} \sigma_1 \\ \sigma_2 \\ \sigma_3 \\ \sigma_4 \\ \sigma_5 \\ \sigma_6 \end{bmatrix} = \begin{bmatrix} C_{11} & C_{12} & C_{13} & C_{14} & C_{15} & C_{16} \\ C_{21} & C_{22} & C_{23} & C_{24} & C_{25} & C_{26} \\ C_{31} & C_{32} & C_{33} & C_{34} & C_{35} & C_{36} \\ C_{41} & C_{42} & C_{43} & C_{44} & C_{45} & C_{46} \\ C_{51} & C_{52} & C_{53} & C_{54} & C_{55} & C_{56} \\ C_{61} & C_{62} & C_{63} & C_{64} & C_{65} & C_{66} \end{bmatrix} \begin{bmatrix} \varepsilon_1 \\ \varepsilon_2 \\ \varepsilon_3 \\ \varepsilon_4 \\ \varepsilon_5 \\ \varepsilon_6 \end{bmatrix}, \tag{2.11}$$

where

$$\begin{bmatrix} \sigma_1 \\ \sigma_2 \\ \sigma_3 \\ \sigma_4 \\ \sigma_5 \\ \sigma_6 \end{bmatrix} = \begin{bmatrix} \sigma_{11} \\ \sigma_{22} \\ \sigma_{33} \\ \sigma_{23} \\ \sigma_{31} \\ \sigma_{12} \end{bmatrix} \quad \text{and} \quad \begin{bmatrix} \varepsilon_1 \\ \varepsilon_2 \\ \varepsilon_3 \\ \varepsilon_4 \\ \varepsilon_5 \\ \varepsilon_6 \end{bmatrix} = \begin{bmatrix} \varepsilon_{11} \\ \varepsilon_{22} \\ \varepsilon_{33} \\ 2\varepsilon_{23} \\ 2\varepsilon_{31} \\ 2\varepsilon_{12} \end{bmatrix}. \tag{2.11a}$$

An elastic homogeneous isotropic medium's physical properties are invariant in all directions. For such a medium, the stiffness tensor $\boldsymbol{C}$ simplifies to

$$\boldsymbol{C} = \begin{bmatrix} \lambda + 2\mu & \lambda & \lambda & 0 & 0 & 0 \\ \lambda & \lambda + 2\mu & \lambda & 0 & 0 & 0 \\ \lambda & \lambda & \lambda + 2\mu & 0 & 0 & 0 \\ 0 & 0 & 0 & \mu & 0 & 0 \\ 0 & 0 & 0 & 0 & \mu & 0 \\ 0 & 0 & 0 & 0 & 0 & \mu \end{bmatrix}. \tag{2.12}$$

Note that there are only two independent elastic moduli in isotropic media. Hooke's law describes the situation of an elastic isotropic medium when it is subjected to small stress, and gives

$$\sigma_{ij} = \lambda \delta_{ij} \varepsilon_{kk} + 2\mu \varepsilon_{ij} = \lambda \delta_{ij} \frac{\partial u_k}{\partial x_k} + \mu \left(\frac{\partial u_i}{\partial x_j} + \frac{\partial u_j}{\partial x_i} \right), \tag{2.13}$$

where a sum is performed over k. In a 2D x–z Cartesian coordinate system,

$$\sigma_{xz} = \mu \left(\frac{\partial u_x}{\partial z} + \frac{\partial u_z}{\partial x} \right). \tag{2.14}$$

$$\sigma_{zz} = \lambda \frac{\partial u_x}{\partial x} + (\lambda + 2\mu)\left(\frac{\partial u_z}{\partial z}\right). \tag{2.15}$$

If stress is persistently exerted, the elastic deformation proceeds to the ductile deformation stage wherein the size and shape changes are irreversible (Nelson 2003). With further exertion of stress, the deformation status reaches the fracture stage where the rock has been broken (Fig. 2.3). At the fracture point, the rock strain is over the strain threshold since real materials are not infinitely rigid.

2.1.4 Stresses and Fractures

Thein-situ stresses can be divided into three perpendicular principal compressive stresses including one vertical stress and two horizontal stresses. Anderson (1951) recognized that the principal stresses orientation ($\boldsymbol{\sigma}_1 > \boldsymbol{\sigma}_2 > \boldsymbol{\sigma}_3$) could vary with geological movement in the upper crust of the earth. Once the three principal stresses deviate from a level of equilibrium, the fractures are possibly created in which the direction of the fracture is parallel to the direction of the maximum stress and perpendicular to the direction of minimum stress. An isotropic medium will be reformed into an anisotropic medium with a new balance of three principal stresses. In fact, in the deeper area of the upper crust, the maximum principal stress is deemed to be produced by compression from the overburden deposition, while the minimum stress will likely be one of the horizontal stresses. Thus most of the fractures in the reservoir are oriented in the vertical or nearly vertical direction because the maximum stress decides the direction of the fracture. Figure 2.4 shows

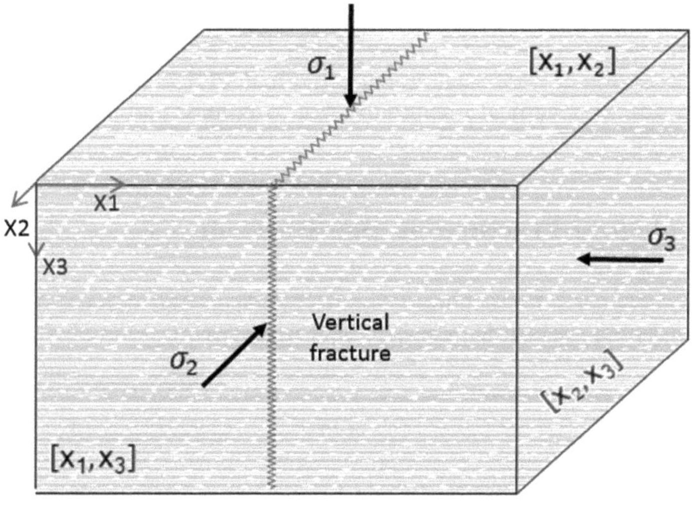

Fig. 2.4 Schematic of vertical fracture with three principal compressive stresses ($\sigma_1 > \sigma_2 > \sigma_3$)

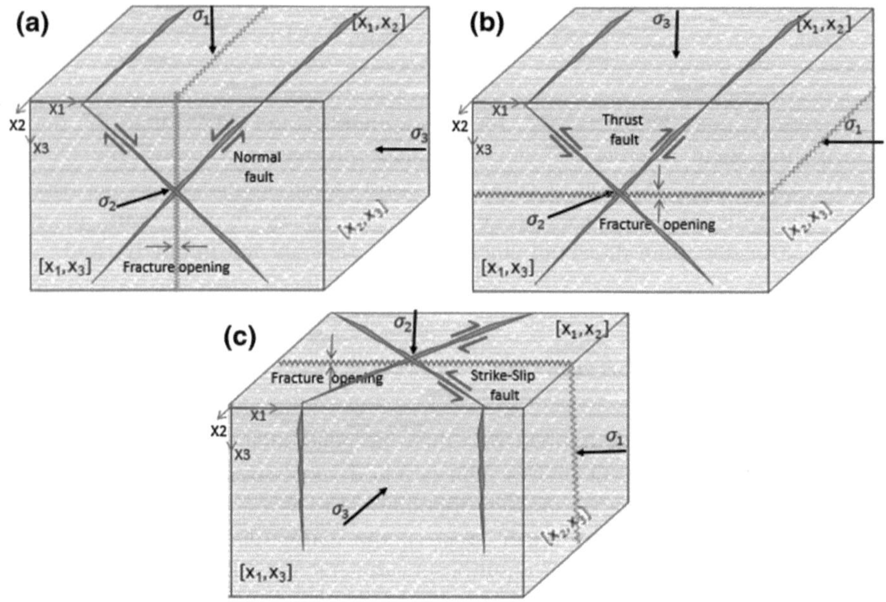

Fig. 2.5 Schematic of the direction relationship between the fractures and the corresponding faults (Anderson 1951) when they are under the same relationship of three principal compressive stresses ordered as $\boldsymbol{\sigma}_1 > \boldsymbol{\sigma}_2 > \boldsymbol{\sigma}_3$

a schematic for three compressive principal stresses and a vertical fracture. If $\boldsymbol{\sigma}_1 > \boldsymbol{\sigma}_2 > \boldsymbol{\sigma}_3$, the vertical fracture plane is parallel to the $[x_2, x_3]$ plane and is perpendicular to the x_1-axis. Figure 2.5 illustrates the direction relationship between the fractures and the corresponding faults (Anderson 1951) when they are under the same relationship of three principal compressive stresses ordered as $\boldsymbol{\sigma}_1 > \boldsymbol{\sigma}_2 > \boldsymbol{\sigma}_3$. Figure 2.5a shows a vertical fracture (normal to x_1-axis) and a normal fault. Figure 2.5b shows a horizontal fracture (normal to x_3-axis) and a thrust fault. Figure 2.5c shows a vertical fracture (normal to x_2-axis) and a strike-slip fault.

2.2 Geological Fractures in the Reservoir

Subsurface geological formationsare confined under in-situ stresses. Therefore, the media in the reservoir can be deformed and fractured.

2.2.1 Fracture Parameters

As we know, the geological fracture is a plane along which rocks show partially lost cohesion when stresses act on it. The fracture planes may exhibit a little displacement or may not exhibit any movement. A slight movement on the fracture plane will produce a fracture opening, which may remain unfilled and result in an increased permeability or may get subsequently filled by secondary minerals or some fluids. Fracture openings, or apertures, and may vary from very thin (0.001–0.5 mm) to relatively wide (1 inch) where fractures are propped open by some minerals in the reservoir (Aguilera 1998).

The subsurface stress is a key to controlling the well-drilling performance and this can be indicated by the fracture orientation. Thus, the most important parameters of a fracture set are the orientation and density which are considered more critical than the fracture opening to describe the fracture sets in a geophysical fracture study. Conventionally, geoscientists describe vertical or nearly vertical fracture orientation that is perpendicular to the direction of minimum compressive stress and the fracture plane as an isotropic plane. The different fractures have different lengths or heights that possibly can span hundreds of meters, and this may result in a lower value of the fracture density in the survey, while it is a fact that there are quite a lot fractures in this area. Fracture density is defined as the number of fractures per meter in a certain direction. For a very sparse fracture set the fracture density could be less than 0.75 m^{-1}, and for a tight fracture set may exceed 10 m^{-1}. Typical values of fracture density for carbonate reservoirs are 1–20 m^{-1} (Bakulin et al. 2000a, b, c). The higher the fracture density is, the higher the permeability is—if the fractures are conductive in the reservoir (Singhal and Gupta 1999).

Figure 2.6 shows two vertical fractures with orientations of φ_1 and φ_2 respect to x_2-axis. It also illustrates the fracture length, fracture height, and fracture opening parameters.

It is clear that fracture aperture is much shorter than seismic wavelengths that are on the order of tens and hundreds of meters. Therefore, in geoscience, a fracture model can effectively be a transversely isotropic model that neglects finite fracture openings, fracture shape, and fracture microstructure. The parameters of such a fractured model will depend on the orientation and intensity of the fracture set(s), as well as on the elastic properties of the host media.

2.2.2 Fracture Detection by Integrated Methods

Fracture detection is an integrated method that incorporates geology, petrophysics, and geophysics technologies. Geological outcrops clearly indicate the near surface fractures from which the directions of local maximum stress can be inferred. Petrophysical well logs directly acquired from the reservoir can delineate the

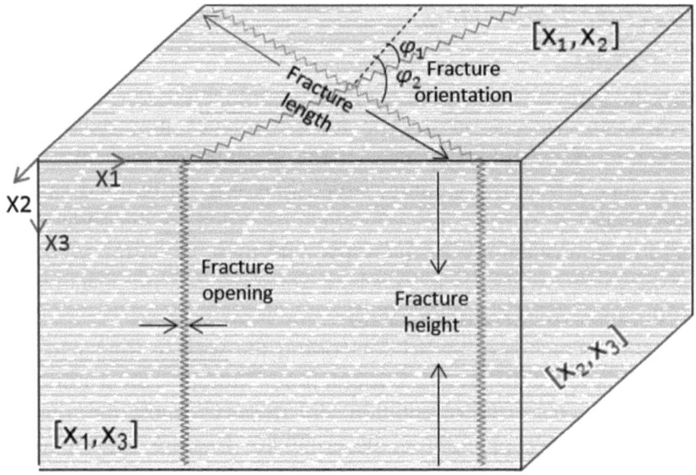

Fig. 2.6 Schematic of two vertical fractures model with parameters of fracture orientation, fracture width and fracture height parameters

fractures through integrated interpretations of logs with a high resolution. Geophysical seismic data provide densely sampled reflection signals from the reservoir that can be used to seek the anisotropic zones indicating possible fractures. In PP-data, COCA gathers (common offset common azimuth), as well as amplitude and velocity variation with offset and azimuth (AVO/VVAZ/AVAZ) data can be used to locate potential fractures. In PS-data, investigation of time lags between fast and slow shear waves is a method to identify the fractures for the reservoirs.

2.2.3 Fracture Delineation by Petrophysical Data

Since the well logs directly measure data from the reservoir, well logs are the most reliable and can be "hard data" for constraining the interpretation of surface seismic data.

2.2.3.1 Core Analysis

The core analysis technique involves the examination of core samples from the zone of interest and performing laboratory examinations. Thus it is the best suited technique for studying fractures on a fine scale and detailed local fracture properties: fracture length and width, the mineralization infilling and un-filling of the fracture, the orientation and density of the fractures, and the contact relationship of the fracture to the host media.

2.2.3.2 Temperature Log

The Earth's temperature gradually increases with depth. A continuous temperature log curve that sharply changes to a cooler temperature can be used to identify the fracture zone. Typically, where the cooler drilling fluid enters into formation, there will be a cool temperature anomaly. A pre-fracturing gradient contrasts with the post-fracturing gradient showing a cool anomaly thought to be the fracture zone (Rider 2002).

2.2.3.3 Caliper Log

Caliper tools measure accurately the hole size, shape, and orientation. A borehole shape named a "breakout" is considered to be an identification for the fracture because breakouts show the orientation of horizontal stress σ_{min} [or a SH_{min} stress (Rider 2002)]. So the detected fractures should be oriented in the direction of maximum horizontal stress σ_{max} [or a SH_{max} stress (Rider 2002)].

2.2.3.4 Density Log

A method of fracture detection is to compare the porosity from the density log with the sonic log. The density tool records the bulk density that includes intergranular and fracture porosities. However, the sonic log just measures intergranular porosity because the sound wave takes the shortest path from emitter to receiver which avoids the fracture. So if the density porosity changes much more than sonic log porosity, it means that fractures are present (Rider 2002).

2.2.3.5 Dipmeter Log

The dipmeter log can show an open fracture from a dipmeter micro-resistivity curve. The indicator is a conductive anomaly due to the invasion of drilling mud. The fracture may be given an orientation because the dipmeter pad bearings are known (Rider 2002).

2.2.3.6 Image Log

An image log is a computer-created image based on acoustic reflectivity or electrical conductivity. To analyze the fracture, it is common to use an acoustic image in which raw acoustic travel times and amplitudes are processed to a color image presentation (Rider 2002).

2.2.4 *Fracture-Induced Anisotropy and Intrinsic Anisotropy*

Anisotropy means that the physical properties of a medium are directionally dependent. Isotropy, as opposed to anisotropy, means that the physical properties of the medium are identical in all directions. In seismology, the anisotropy properties commonly can be classified into intrinsic anisotropy and induced anisotropy.

The intrinsic anisotropy may be caused in the original formation in sedimentary zones when the layers are substantially thinner than the seismic wavelength. It assumes that the layers have a welded contact and the waves should meet the perfectly welded boundary conditions. Intrinsic anisotropy is usually described by Thomsen's anisotropic parameters (1986).

Induced anisotropy appears because a medium is subject to regional stresses that cause the medium to crack and attain a new system of equilibrium stresses in the subsurface. The cracking action causes a welded continuous medium to change to a nonwelded discontinuous medium satisfying the nonwelded contact boundary conditions. From theoretical work and field observations, the fractured medium exhibits induced anisotropy (Crampin and Bamford 1977; Lefeuvre 1993; Lynn et al. 1996; Ramos and Davis 1997; Rueger 1996). The induced anisotropy shows azimuthal dependence that is visible as a sinusoidal variation in the seismic data in which the travel time is a function of the azimuth (due to velocity variation) in the common offset and common azimuth cube (COCA). The shear wave splitting phenomenon further indicates the azimuthal dependence of the induced anisotropy in that the shear wave split into a fast shear wave that the polarization is parallel to the direction of the fracture, and into a slower shear wave that the polarization is perpendicular the direction of the fracture. Both polarizations are orthogonal to the wave propagation direction (Crampin 1985).

The fracture causes the induced anisotropy that can be described by the fracture weakness parameters that appear in the five independent moduli of the fractured medium in next Sect. (2.3).

2.2.4.1 Anisotropic Parameters and Stiffness

In 1986, Thomsen introduced three dimensionless weak-anisotropic parameters: $\varepsilon_{\mathrm{Th}}, \delta_{\mathrm{Th}}$, and γ_{Th} that are related to the five independent moduli for TI media as

$$\gamma_{\mathrm{Th}} \equiv \frac{C_{66} - C_{44}}{2C_{44}}, \tag{2.16a}$$

$$\varepsilon_{\mathrm{Th}} \equiv \frac{C_{11} - C_{33}}{2C_{33}}, \tag{2.16b}$$

$$\delta_{\mathrm{Th}} \equiv \frac{(C_{13} + C_{44})^2 - (C_{33} - C_{44})^2}{2C_{33}(C_{33} - C_{44})}, \tag{2.16c}$$

also,

$$\alpha_{p0} \equiv \sqrt{C_{33}/\rho}, \tag{2.16d}$$

$$\beta_{p0} \equiv \sqrt{C_{44}/\rho}, \tag{2.16e}$$

where γ_{Th} describes the fractional difference between the horizontal and vertical SH-wave velocities, $\varepsilon_{\mathrm{Th}}$ describes the fractional difference between the horizontal and vertical P-wave velocities, and δ_{Th} denotes the variation of P-wave velocity with phase angle for nearly vertical propagation (Thomsen 1986). α_{p0} and β_{p0} are vertical P-wave and S-wave velocities, respectively.

Based on the definitions of the weak-anisotropy coefficients in Eqs. (2.16a–2.16e), the notations for the VTI (transversely isotropic medium with vertical symmetric axis) and HTI (transversely isotropic medium with horizontal symmetric axis) anisotropy coefficients are often identified as $\gamma, \varepsilon, \delta$, and $\gamma^{(v)}, \varepsilon^{(v)}, \delta^{(v)}$ respectively, and they have been widely used in the geophysical literature to describe wave propagation in anisotropic media.

Tsvankin (1997a, b) introduced the anisotropic coefficients for an orthorhombic medium with the superscript notations (1), (2) and (3) for three-dimensional symmetric planes using Thomson's-type weak-anisotropic parameters (Fig. 2.7). The superscript indicates the direction of the axis that is normal to the symmetry

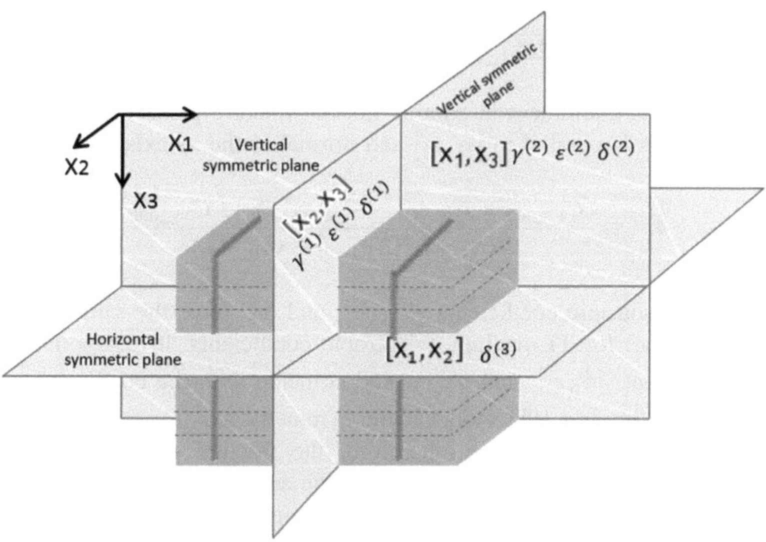

Fig. 2.7 Sketch of an orthorhombic media model composed of two vertical fractures embedded in a layered host medium (VTI). Two vertical symmetric planes and one horizontal symmetric plane are determined by the vertical fracture orientation and the horizontal layered medium

plane. For example, the notation (1) means that the x_1-axis is normal to the symmetry plane.

The weak-anisotropy coefficients in the fracture plane (a vertical symmetric plane) of the orthorhombic medium that is parallel to $[x_2, x_3]$ and normal to the x_1-axis are

$$\gamma^{(1)} \equiv \frac{C_{66\_orth} - C_{55\_orth}}{2C_{55\_orth}}, \tag{2.17a}$$

$$\varepsilon^{(1)} \equiv \frac{C_{22\_orth} - C_{33\_orth}}{2C_{33\_orth}}, \tag{2.17b}$$

$$\delta^{(1)} \equiv \frac{(C_{23\,orth} + C_{44_{orth}})^2 - (C_{33_{orth}} - C_{44_{orth}})^2}{2C_{33_{orth}}(C_{33_{orth}} - C_{44_{orth}})}. \tag{2.17c}$$

The weak-anisotropy coefficients in the normal fracture plane (another vertical symmetric plane, see Table 2.2) that is parallel to $[x_1, x_3]$ and normal to the x_2-axis are

$$\gamma^{(2)} \equiv \frac{C_{66\_orth} - C_{44\_orth}}{2C_{44\_orth}}, \tag{2.18a}$$

$$\varepsilon^{(2)} \equiv \frac{C_{11\_orth} - C_{33\_orth}}{2C_{33\_orth}}, \tag{2.18b}$$

$$\delta^{(2)} \equiv \frac{(C_{13\_orth} + C_{55\_orth})^2 - (C_{33\_orth} - C_{55\_orth})^2}{2C_{33\_orth}(C_{33\_orth} - C_{55\_orth})}. \tag{2.18c}$$

The anisotropic coefficient in another normal fracture plane (a horizontal symmetric plane) that is parallel to $[x_1, x_2]$ and normal to the x_3 axis is

$$\delta^{(3)} \equiv \frac{(C_{12\_orth} + C_{66\_orth})^2 - (C_{11\_orth} - C_{66\_orth})^2}{2C_{11\_orth}(C_{11\_orth} - C_{66\_orth})}. \tag{2.19}$$

Note that anisotropic coefficients $\gamma^{(2)}, \varepsilon^{(2)}$, and $\delta^{(2)}$ have the same formulae as Eqs. (2.16a–2.16c) for TI medium anisotropic coefficients. It means that the anisotropic coefficient $\gamma^{(2)}, \varepsilon^{(2)}$ and $\delta^{(2)}$ would coincide with the coefficients $\gamma^{(v)}, \varepsilon^{(v)}$ and $\delta^{(v)}$ for the simplest HTI model if the symmetry axis is orientated in the x_1-direction. In the isotropy plane, therefore, the fracture does not affect wave velocities and the propagation direction, and the anisotropic coefficients will have the same constant values as the host VTI anisotropic medium (Tsvankin 1997a).

For fractured media, when the fractures have a certain direction (one or several), and the wavelengths are much greater than the fracture opening, it is feasible that the induced anisotropic problem described by the fracture weakness parameters can be transformed into the intrinsic anisotropic problem described by Thomson's anisotropy parameters with Tsvankin's notations (see Sect. 2.3).

2.3 Fracture Related Geophysical Assumptions

Relative to the seismic wavelength, a fracture is regarded as a weakness plane. Thus, a seismological fractured medium model should neglect finite fracture openings, fracture microstructure, and it can be equivalent to an effective anisotropic medium.

2.3.1 *Backus Average Theory*

Backus (1962) presented a long wavelength equivalent theory that describes a finely stratified homogeneous medium which is effectively equivalent to an anisotropic medium, and is named an "effective anisotropic medium". The effective anisotropic medium might be a transversely isotropic medium with a vertical symmetry axis (VTI) or transversely isotropic medium with a horizontal symmetry axis (HTI). In Fig. 2.8, there is a picture of Backus's theory (1962): the thin layers are parallel to the horizontal x_1-axis and the media properties vary with vertical x_3-axis. The medium thickness H must be long enough so that the elastic properties of the medium vary appreciably over H. Also it must be smaller than the smallest wavelength in order to replace the layered medium by an anisotropic medium where the density is the average density over H and the elastic parameters are an algebraic

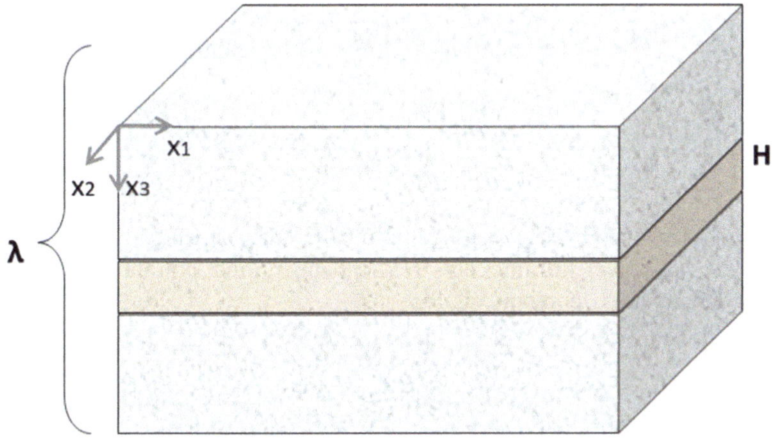

Fig. 2.8 Sketch of the long wavelength equivalent medium. H is the medium width.λ is seismic wavelength.$\lambda \gg$ **H**

combination of the parameters from the original layered medium. The theory concludes that the elastic moduli of the equivalent medium can be expressed, in the long wavelength assumption, as thickness-weighted averages of the moduli of the thin layers of the stratified medium. The Backus average theory has been verified numerically by Carcione et al (1991).

2.3.2 Stress, Strain in the Stratified Layers

For a stratified medium, in the long wavelength limit, all components of stress acting on the layering plane, i.e., σ_{3h}, σ_{4h} and σ_{5h} ($h = 1 \ldots n$, with h denoting different layers) and all components of strain lying in the layering plane, i.e., ε_{1h}, ε_{2h} and ε_{6h}, are the same in all the layers across medium H. The notations of stress and strain are in Voigt form. The other components of the stress and strain, i.e., $\sigma_{1h}, \sigma_{2h}, \sigma_{6h}$ and $\varepsilon_{3h}, \varepsilon_{4h}, \varepsilon_{5h}$, are different from layer to layer. In other words, some components of stress and strain are layer-independent, i.e.,

$$\begin{bmatrix} \sigma_{3h} \\ \sigma_{4h} \\ \sigma_{5h} \end{bmatrix} = \begin{bmatrix} \sigma_3 \\ \sigma_4 \\ \sigma_5 \end{bmatrix} = \sigma, \tag{2.20a}$$

$$\begin{bmatrix} \varepsilon_{1h} \\ \varepsilon_{2h} \\ \varepsilon_{6h} \end{bmatrix} = \begin{bmatrix} \varepsilon_1 \\ \varepsilon_2 \\ \varepsilon_6 \end{bmatrix} = \varepsilon, \tag{2.20b}$$

whereas other components of stress and strain are layer-dependent, i.e.,

$$\begin{bmatrix} \sigma_{1h} \\ \sigma_{2h} \\ \sigma_{6h} \end{bmatrix} = \sigma_h, \tag{2.21a}$$

$$\begin{bmatrix} \varepsilon_{3h} \\ \varepsilon_{4h} \\ \varepsilon_{5h} \end{bmatrix} = \varepsilon_h. \tag{2.21b}$$

The thickness-weighted average value for the layer-dependent components will be the total value over full thickness H (Schoenberg and Douma 1988). Thus, the layer-dependent stress components are

$$\overline{\sigma}_1 = \frac{1}{H} \sum_{h=1}^{n} H_h \sigma_{1h}, \tag{2.22a}$$

$$\overline{\sigma}_2 = \frac{1}{H}\sum_{h=1}^{n} H_h \sigma_{2h}, \tag{2.22b}$$

$$\overline{\sigma}_6 = \frac{1}{H}\sum_{h=1}^{n} H_h \sigma_{6h}, \tag{2.22c}$$

the layer-dependent strain components are

$$\overline{\varepsilon}_3 = \frac{1}{H}\sum_{h=1}^{n} H_h \varepsilon_{3h}, \tag{2.23a}$$

$$\overline{\varepsilon}_4 = \frac{1}{H}\sum_{h=1}^{n} H_h \varepsilon_{4h}, \tag{2.23b}$$

$$\overline{\varepsilon}_5 = \frac{1}{H}\sum_{h=1}^{n} H_h \varepsilon_{5h}, \tag{2.23c}$$

where H_h is the thickness and l_n is the relative thickness of the composed thin layer, and $l_1 + l_2 + \cdots + l_n = 1$. In Eqs. (2.22a) and (2.23a), the overhead bar denotes that the layer-dependent components of stress and strain have been done using the thickness-weighted average. The linear relationship of the stress and strain in Eq. (2.11) can be separated into two parts that are individually for the layer-dependent and layer-independent stress components with corresponding strain components and the thickness-weighted average for the stiffness

$$\overline{\boldsymbol{\sigma}} = \boldsymbol{C}_{\mathbf{TT}}\boldsymbol{\varepsilon} + \boldsymbol{C}_{\mathbf{TN}}\overline{\boldsymbol{\varepsilon}}, \tag{2.24a}$$

$$\boldsymbol{\sigma} = \boldsymbol{C}_{\mathbf{NT}}\boldsymbol{\varepsilon} + \boldsymbol{C}_{\mathbf{NN}}\overline{\boldsymbol{\varepsilon}}, \tag{2.24b}$$

where $\boldsymbol{C}_{\mathbf{TN}}$ is the transpose of the corresponding $\boldsymbol{C}_{\mathbf{NT}}$, and

$$\boldsymbol{C}_{\mathbf{TT}} = \begin{bmatrix} C_{11} & C_{12} & C_{16} \\ C_{21} & C_{22} & C_{26} \\ C_{16} & C_{26} & C_{66} \end{bmatrix}, \tag{2.25a}$$

$$\boldsymbol{C}_{\mathbf{NN}} = \begin{bmatrix} C_{33} & C_{34} & C_{35} \\ C_{34} & C_{44} & C_{45} \\ C_{35} & C_{45} & C_{55} \end{bmatrix}, \tag{2.25b}$$

$$\boldsymbol{C}_{\mathbf{TN}} = \begin{bmatrix} C_{13} & C_{14} & C_{15} \\ C_{23} & C_{24} & C_{25} \\ C_{36} & C_{46} & C_{56} \end{bmatrix}. \tag{2.25c}$$

2.3.3 *Stiffness and Compliances of the Fracture*

In an effective medium, when a thin layer composing the medium is permanently deformed or fractured, this layer will be soft, and the layer-dependent strain components $\bar{\boldsymbol{\varepsilon}}$ will be enlarged. The layer-independent strain components $\boldsymbol{\varepsilon}$ are the same as the corresponding components in the host medium with a constant value. To represent the fracture, Eq. (2.24b) can be approximated as

$$\boldsymbol{\sigma}_f \approx \boldsymbol{C}_{\mathrm{NN}f}\bar{\boldsymbol{\varepsilon}}. \tag{2.26}$$

Equation (2.26) describes a pure fracture feature. It turns out that the fracture feature is mainly affected by the stiffness $\boldsymbol{C}_{\mathbf{NN}f}$ rather than by the stiffness $\boldsymbol{C}_{\mathbf{TT}}$ and $\boldsymbol{C}_{\mathbf{TN}}$. $\boldsymbol{C}_{\mathbf{NN}f}$ is approximately linearly proportional to the layer-independent stresses and layer-dependent strains. Now it is assumed that the fracture plane is isotropic (no roughness) and the fracture behavior is invariant with rotation with respect to an axis normal to the fracture (fracture system), and the fracture compliance is $\boldsymbol{S}$ (Schoenberg 1988, 1995). The fracture stiffness $\boldsymbol{C}_{\mathbf{NN}f}$ and fracture compliance $\boldsymbol{S}$ can be transformed into each other as follows:

$$\begin{aligned}\boldsymbol{C}_{\mathrm{NN}f}^{-1} &= \begin{bmatrix} C_{33} & 0 & 0 \\ 0 & C_{44} & 0 \\ 0 & 0 & C_{55} \end{bmatrix}^{-1} \\ &= \boldsymbol{S} = \begin{bmatrix} S_{\mathrm{N}} & 0 & 0 \\ 0 & S_H & 0 \\ 0 & 0 & S_V \end{bmatrix},\end{aligned} \tag{2.27}$$

where S_{N} is the normal fracture compliance that relates to the normal displacement and the normal stress, S_V and S_H are the tangential fracture compliances along the vertical (x_3-axis) and horizontal (x_1-axis) axes, respectively. Equation (2.27) is the so-called rotation-invariant fracture system. Because the two tangential fracture compliances are equal in the rotation-invariant fracture system, S_{T} may be used ($S_{\mathrm{T}} = S_V = S_H$) as the tangential compliance that relates to the shear displacement and the shear stress. The fracture compliances S_{N} and S_{T} are nonnegative and have the dimension length/stress within a discontinuous medium.

In 1988, Schoenberg and Douma (1988) presented the dimensionless fracture compliances that link the dimensioned fracture compliances to the unfractured host medium compliances:

$$E_{\mathrm{T}} \equiv C_{44H}S_{\mathrm{T}}, \tag{2.28a}$$

$$E_{\mathrm{N}} \equiv C_{33H}S_{\mathrm{N}}, \tag{2.28b}$$

where E_N and E_T are the normal and tangential dimensionless fracture compliances. The subscript "*H*" denotes the host medium. To simplify the fracture compliances, Hsu and Schoenberg (1993) introduce the dimensionless quantities

$$\Delta_N = \frac{E_N}{1+E_N}, \tag{2.29a}$$

$$\Delta_T = \frac{E_T}{1+E_T}, \tag{2.29b}$$

$$0 \leq \Delta_N, \quad \Delta_T \leq 1. \tag{2.29c}$$

The quantities Δ_N, Δ_T relate the fracture compliance to the total compliance of a fractured medium that is algebraically a summation of a host medium and a fracture. Δ_N, Δ_T are also the so-called normal and tangential weakness of the fracture when the fracture is treated as an infinite slip interface (Bakulin et al. 2000a, b, c).

2.3.4 Linear Slip Interface and Fracture

A linear slip interface can be used to model a fracture that is based on the Backus average (1962). In 1980, Michael Schoenberg depicted a physical mechanism of the linear slip interface: an isotropic thin layer inserted into a homogeneous isotropic host medium to construct a perfectly welded layered medium (effective anisotropic VTI media). This isotropic thin layer thickness l' should be very small compared to a minimum wavelength λ, and its impedance Z' should be much lower with respect to the host medium impedance Z. The reflection and transmission coefficients of this isotropic thin layer are R' and T'. Once the thickness and impedance of the inserted isotropic thin layer approach to zero, the constructed layered medium (Fig. 2.9) is transformed into a medium that is effectively equal to a linear slip interface embedded in a homogeneous isotropic host medium in which the displacement discontinuity and stress continuity across the linear slip interface satisfy the nonwelded contact boundary conditions (Schoenberg 1980). At this point, the transformed medium is named the effective fractured medium, and the slip interface equals to a fracture. The reflection coefficient R and transmission coefficient T of the effective fractured medium will be $R = R'$ and $T = T'$.

The linear slip theory appreciably provides a convenient effective fractured medium model in geophysics that allows us to study the fracture from seismic signatures since the fracture is possibly simulated as an infinite weakness surface. Pyrak-Nolte et al. (1990a, b) and Hsu and Schoenberg (1993) have confirmed the validity of this model by some laboratory experimental verifications. In addition, Peterson et al. (1993) have shown that results obtained from small-scale cross-well experiments appear to agree with this model. In fact, the fracture outcrops show that a fracture as an nonwelded interface or weakness plane separates a medium into two

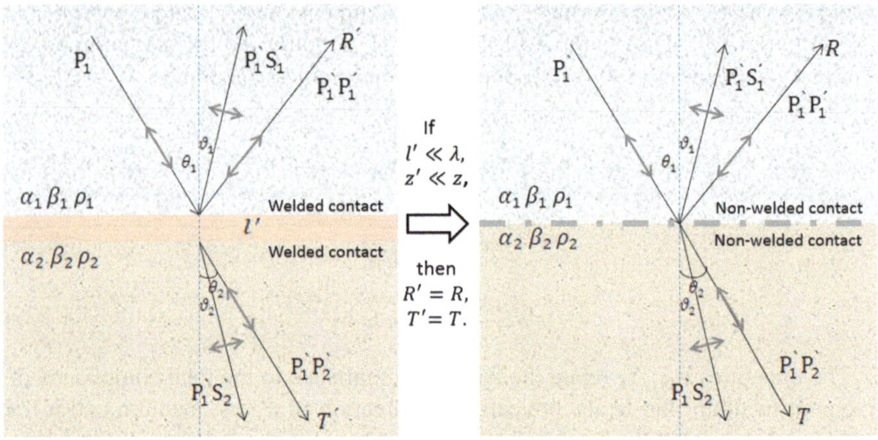

Fig. 2.9 Diagram of physical mechanism of the linear slip interface. Once $l' << \lambda$, $z' << z$, then $\boldsymbol{R}' = \boldsymbol{R}$, $\boldsymbol{T}' = \boldsymbol{T}$

half spaces that exactly coincide with the linear slip interface. The linear slip interface theory satisfies the nonwelded contact boundary conditions. Therefore, a linear slip interface obviously can be adapted to simulate a fracture which unites a geometrical phenomenon of the geology and a mechanical property of the medium. In the effective fractured medium model, the fracture is just an infinitely extended weakness plane or a highly compliant layer, regardless of its shape and microstructure and the porosity information.

Sayers (2009) used the ratio between the normal and tangential compliances as an indicator for the fluid content, e.g., $\frac{S_N}{S_T} \approx 1$ for dry fractures. Schoenberg (1980) suggested that a fluid filled fracture can be approximated by letting $S_N = 0$ and $S_T \neq 0$, which is equivalent to requiring normal displacement continuity. The case of $S_N = 0$ and $S_T \propto \infty$, meaning that there is no shear stress across the interface, and is equivalent to the solid–fluid boundary condition: both the normal stress and normal displacement are continuous.

2.3.5 Schoenberg-Muir Calculus Theory

Schoenberg and Muir (1989) extended the Backus average approach to develop a matrix formalism that enables a simple calculation of the stiffness of the rock, and then achieves a description for the composition and decomposition of the effective anisotropic medium. First, the rock stiffness of a layer can be mapped to an element of an Abelian group that includes two scalars and three 3×3 matrices. Then summation or subtraction is used to calculate the corresponding elements from the media that need to be calculated. Finally, the summed or the subtracted elements are inverted into the rock stiffness to describe a reconstructed medium. Under the Backus theory (1962), a numerical simulation was performed to show that the Schoenberg-Muir theory is valid from the kinematic (travel times) and dynamic (amplitudes) viewpoints for a small crack aspect ratio or fracture opening, very long flat parallel fractures and thin layered media (Carcione et al. 2012). It concludes that a fracture as an infinitely extended weakness plane is an element for assembling a fractured medium from a fracture and a host medium. A fractured medium can be separated into a fracture and a host medium

$$\text{Fracture} + \text{Host medium} \rightarrow \text{Fractured medium}$$
$$\text{Fractured medium} \rightarrow \text{Fracture} + \text{Host medium.}$$

In this view, geophysically a horizontally fractured medium (the VTI case) can be composed as a horizontal slip interface (a horizontal fracture) within an isotropic homogeneous host medium. A vertically fractured mediumconsists of a vertical linear slip interface (a vertical fracture) and an isotropic homogeneous host medium. A more realistic orthorhombic medium can be formed by either embedding a vertical slip interface into a transversely isotropic host medium with a symmetric vertical axis (VTI) or embedding a horizontal slip interface into a transversely isotropic host medium with a symmetric horizontal axis (HTI).

Schoenberg and Muir's calculus approach has been further developed into an algorithm that calculates the rock compliances instead of the rock stiffnesses by Hood (1991) and Nichols et al (1989).

Table 2.1 gives the types of media contact that will be involved in this book. Throughout the book, the expression "a fractured medium" means that the host medium is a uniform homogeneous isotropic medium, while the expression "fractured medium with impedance contrast" implies that homogeneous isotropic media with impedance contrasts are the host media.

2.3.6 *Horizontally Fractured Medium Moduli (VTI)*

The stiffness matrix of an isotropic medium (Eq. 2.12) can be rewritten in a special order of the elements by moving the column six into column three and then moving the row six into row three as

$$C_{H\_\text{iso}} = \begin{bmatrix} C_{11} & C_{12} & 0 & C_{13} & 0 & 0 \\ C_{21} & C_{22} & 0 & C_{23} & 0 & 0 \\ 0 & 0 & C_{66} & 0 & 0 & 0 \\ C_{31} & C_{32} & 0 & C_{33} & 0 & 0 \\ 0 & 0 & 0 & 0 & C_{44} & 0 \\ 0 & 0 & 0 & 0 & 0 & C_{55} \end{bmatrix}. \tag{2.30}$$

Table 2.1 Types of media contact

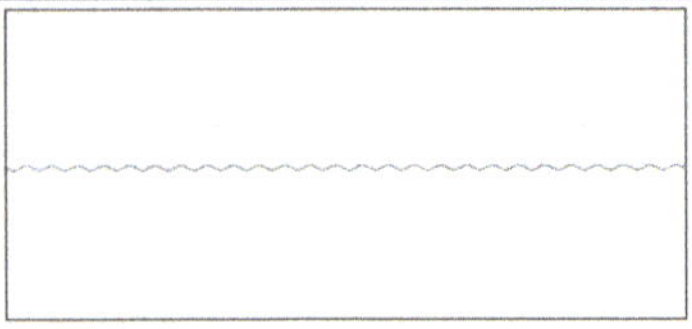

Fracture interface

Impedance contrast interface

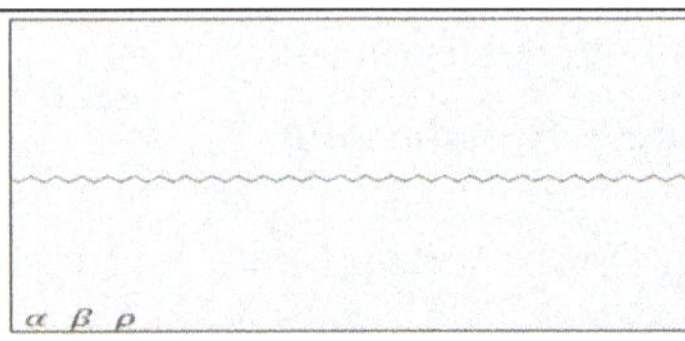

Horizontally fractured medium

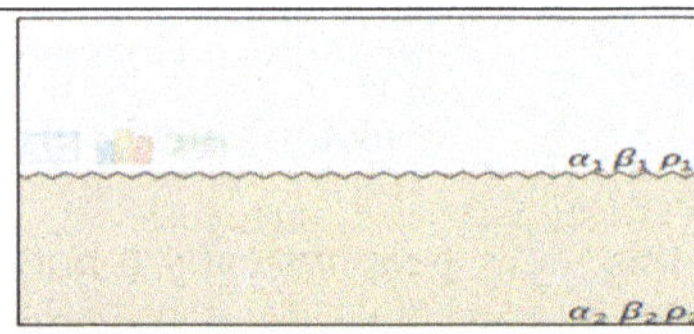

Horizontally fractured medium with impedance contrast

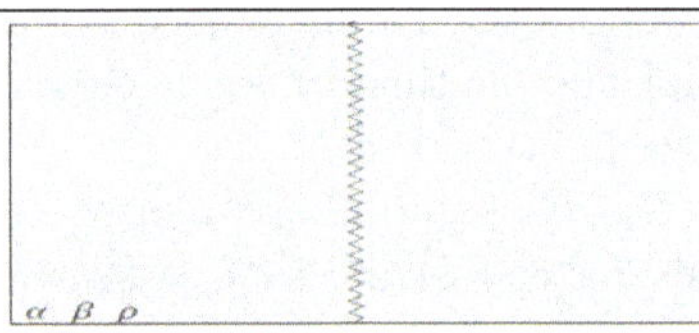

Vertically fractured medium

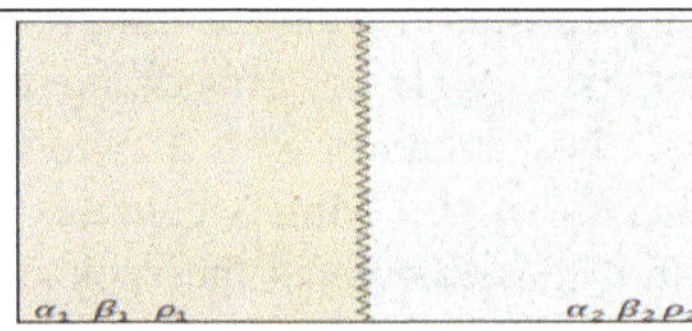

Vertically fractured medium with impedance contrast

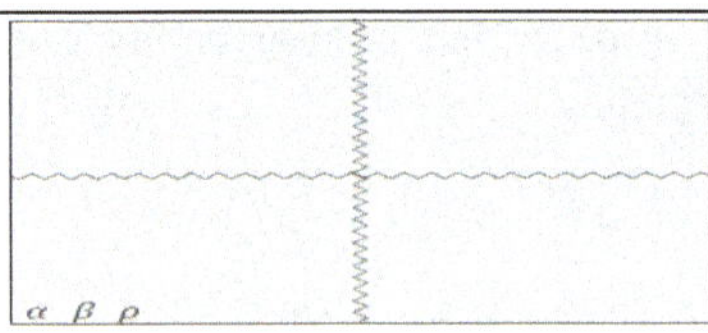

Orthogonally fractured medium

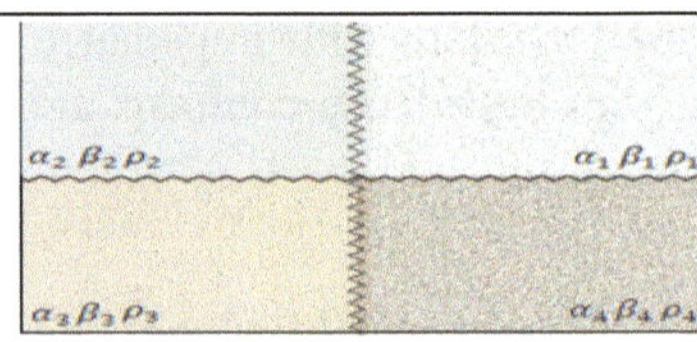

Orthogonally fractured medium with impedance contrast

Note that the nine elements of the top left of the matrix in Eq. (2.30) form the stiffness matrix $\boldsymbol{C}_{TT}$, while the nine elements (3×3) of the bottom right of the matrix in Eq. (2.30) form the stiffness matrix $\boldsymbol{C}_{NN}$ (Appendix A). Following the matrix formalism of algebra and Abelian (commutative) group theory (Schoenberg and Muir 1989), the effective elastic moduli of the fractured VTI media (Fig. 2.10) can be computed as

Fig. 2.10 Horizontally fractured medium model. It is formed by a horizontal fracture interface and a uniform homogeneous isotropic host medium

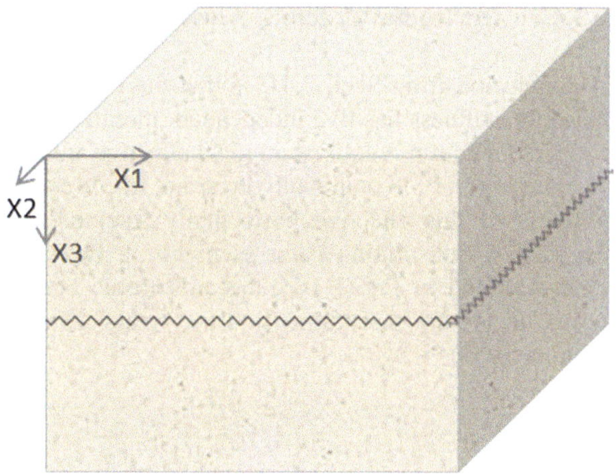

$$\boldsymbol{G}_{\mathbf{VTI}} = \boldsymbol{G}_{\boldsymbol{H}\_\mathbf{iso}} + \boldsymbol{G}_{\boldsymbol{h}\_\boldsymbol{f}}, \tag{2.31}$$

where $\boldsymbol{G}_{\boldsymbol{H}\_\mathbf{iso}}$ the group stiffness of the homogeneous isotropic host medium, and $\boldsymbol{G}_{\boldsymbol{h}\_\boldsymbol{f}}$ is the group stiffness of the horizontal fracture that was discussed and given in Eq. (2.27). The effective elastic stiffness $\boldsymbol{C}_{\mathbf{VTI}}$ of the horizontally fractured media with the parameters of tangential and normal weaknesses of the fracture (Eqs. 2.29a) is (see Appendix A)

$$\boldsymbol{C}_{\mathbf{VTI}} = \begin{bmatrix} (\lambda+2\mu)\left[1-\frac{\lambda^2}{(\lambda+2\mu)^2}\Delta_{\mathrm{N}}\right] & \lambda\left(1-\frac{\lambda}{\lambda+2\mu}\Delta_{\mathrm{N}}\right) & \lambda(1-\Delta_{\mathrm{N}}) & 0 & 0 & 0 \\ \lambda\left(1-\frac{\lambda}{\lambda+2\mu}\Delta_{\mathrm{N}}\right) & (\lambda+2\mu)\left[1-\frac{\lambda^2}{(\lambda+2\mu)^2}\Delta_{\mathrm{N}}\right] & \lambda(1-\Delta_{\mathrm{N}}) & 0 & 0 & 0 \\ \lambda(1-\Delta_{\mathrm{N}}) & \lambda(1-\Delta_{\mathrm{N}}) & (\lambda+2\mu)(1-\Delta_{\mathrm{N}}) & 0 & 0 & 0 \\ 0 & 0 & 0 & \mu(1-\Delta_H) & 0 & 0 \\ 0 & 0 & 0 & 0 & \mu(1-\Delta_V) & 0 \\ 0 & 0 & 0 & 0 & 0 & \mu \end{bmatrix}. \tag{2.32}$$

This fractured elastic stiffness shows that the fractured medium is a TI medium with five independent moduli $(C_{\mathrm{VTI}12} = C_{\mathrm{VTI}11} - C_{\mathrm{VTI}66})$ and the fracture properties are independent of the properties of the host medium (Hsu 1993).

2.3.6.1 Horizontal Fracture Anisotropy

A conclusion from Sect. 2.4.2 is that the effective composite media are TI media since the stiffness has five independent moduli. Hereby, the composed horizontally fractured medium is equivalent to an effective anisotropic VTI medium in which the fracture is part of a rotationally invariant fracture system and causes anisotropy. The stiffness of this effective horizontally fractured medium has been presented in Eq. (2.32). Substituting some elements in Eq. (2.32) into the anisotropic coefficients defined in Eq. (2.16a, the anisotropic coefficients caused by a fracture in terms of the fracture dimensionless weakness parameters are (Schoenberg and Douma 1988)

$$\gamma = \frac{\Delta_{\mathrm{T}}}{2(1-\delta_{\mathrm{T}})} \approx \frac{\Delta_{\mathrm{T}}}{2}, \tag{2.33a}$$

$$\varepsilon = \frac{2g(1-g)\Delta_{\mathrm{N}}}{1-\delta_{\mathrm{N}}} \approx 2g(1-g)\Delta_{\mathrm{N}}, \tag{2.33b}$$

$$\delta = 2(1-g)\frac{g(\Delta_{\mathrm{N}}-\Delta_{\mathrm{T}})}{(1-\Delta_{\mathrm{N}})-g(1-\Delta_{\mathrm{T}})} \approx 2(1-g)g(\Delta_{\mathrm{N}}-\Delta_{\mathrm{T}}), \tag{2.33c}$$

where g is

$$g = \frac{\mu_H}{\lambda_H + 2\mu_H}. \tag{2.33d}$$

2.3.7 *Vertically Fractured Medium Moduli (HTI)*

Cui and Lines (2011) showed that the elastic moduli of the HTI can be computed from the elastic moduli of a transversely isotropic medium with a vertical symmetry axis (a VTI medium) by using a Bond transformation (Winterstein 1990). Hence, rotating the horizontal fracture system in Eq. (2.27) by 90° with respect to the Y axis (Fig. 2.11), the elastic moduli of the vertical fracture system should be

$$\begin{aligned} \boldsymbol{S}_{v\_f} &= \begin{bmatrix} C_{11} & 0 & 0 \\ 0 & C_{55} & 0 \\ 0 & 0 & C_{66} \end{bmatrix}^{-1} \\ &= \begin{bmatrix} S_N & 0 & 0 \\ 0 & S_V & 0 \\ 0 & 0 & S_H \end{bmatrix}. \end{aligned} \tag{2.34}$$

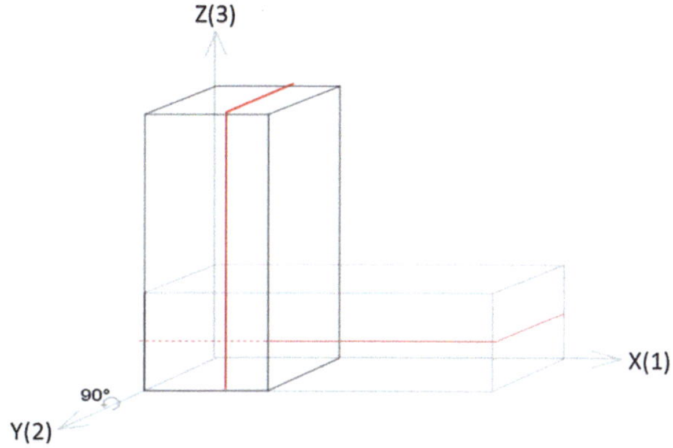

Fig. 2.11 Coordinates Rotation. A horizontal interface is rotated into a vertical interface by a rotation of 90° with respect to the *Y* axis

Figure 2.11 shows a vertical fracture that is normal to the $X(1)$-axis or parallel to $[Y(2), Z(3)]$-plane with a horizontal symmetry axis. It is rotated from a horizontal fracture.

Following the Schoenberg and Muir (1989) procedure to simplify the calculation of the effective elastic moduli of the vertically fractured media (HTI) (Fig. 2.12), we have:

$$\boldsymbol{G}_{\mathbf{HTI}} = \boldsymbol{G}_{H\_\mathbf{iso}} + \boldsymbol{G}_{v\_f}. \tag{2.35}$$

$\boldsymbol{G}_{H\_\mathbf{iso}}$ is the group stiffness of the homogeneous isotropic host medium, $\boldsymbol{G}_{v\_f}$ is the group stiffness of the vertical fracture. Consequently, the elastic stiffness for a vertically fractured medium (HTI) is (Appendix A)

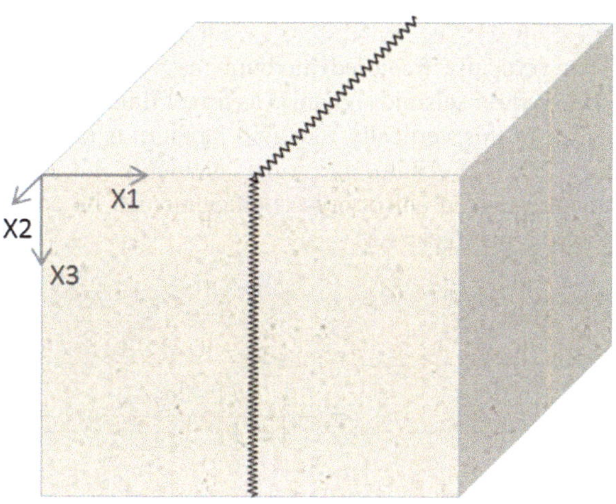

Fig. 2.12 Vertically fractured medium model. It is formed by a vertical fracture interface and a uniform homogeneous isotropic host medium

$$\boldsymbol{C}_{\mathbf{HTI}} = \begin{bmatrix} (\lambda+2\mu)(1-\Delta_{\mathrm{N}}) & \lambda(1-\Delta_{\mathrm{N}}) & \lambda(1-\Delta_{\mathrm{N}}) & 0 & 0 & 0 \\ \lambda(1-\Delta_{\mathrm{N}}) & (\lambda+2\mu)\left[1-\frac{\lambda^2}{(\lambda+2\mu)^2}\Delta_{\mathrm{N}}\right] & \lambda\left(1-\frac{\lambda}{\lambda+2\mu}\Delta_{\mathrm{N}}\right) & 0 & 0 & 0 \\ \lambda(1-\Delta_{\mathrm{N}}) & \lambda\left(1-\frac{\lambda}{\lambda+2\mu}\Delta_{\mathrm{N}}\right) & (\lambda+2\mu)\left[1-\frac{\lambda^2}{(\lambda+2\mu)^2}\Delta_{\mathrm{N}}\right] & 0 & 0 & 0 \\ 0 & 0 & 0 & \mu & 0 & 0 \\ 0 & 0 & 0 & 0 & \mu(1-\Delta_V) & 0 \\ 0 & 0 & 0 & 0 & 0 & \mu(1-\Delta_H) \end{bmatrix}. \tag{2.36}$$

The elements of Eq. (2.36) have $C_{\mathrm{HTI22}} = C_{\mathrm{HTI33}}, C_{\mathrm{HTI12}} = C_{\mathrm{HTI13}}, C_{\mathrm{HTI55}} = C_{\mathrm{HTI66}}$ and $C_{\mathrm{HTI23}} = C_{\mathrm{HTI33}} - 2C_{\mathrm{HTI44}}$. Thus effective vertically fractured media are TI media with five independent moduli. Note that $C_{44\_\mathrm{HTI}} = \mu$ is the same as the homogeneous isotropic host medium stiffness. It means that only the elements corresponding to the fracture system describe the fracture in the stiffness for the effective vertically fractured medium. This effective vertically fractured medium (HTI) is the simplest azimuthally anisotropic medium that provides valuable insights into the study of the fracture behavior of the seismic signature variation with the azimuthal fractures.

2.3.7.1 Vertical Fracture Anisotropy

This vertically fractured medium may cause anisotropy that would be seen and observed on seismic signature as travel time and amplitude variations. The stiffness tensor of this vertically fractured medium is expressed by the fracture weakness in Eq. (2.36). Substituting some elements in Eq. (2.36) into Eq. (2.16a), the fracture-caused anisotropic coefficients can be conveniently described by the anisotropic parameters as

$$\gamma^{(v)} = -\frac{\Delta_{\mathrm{T}}}{2}, \tag{2.37a}$$

$$\varepsilon^{(v)} = -\frac{2g(1-g)\Delta_{\mathrm{N}}}{1-(1-2g)^2\Delta_{\mathrm{N}}} \approx -2g(1-g)\Delta_{\mathrm{N}}, \tag{2.37b}$$

$$\delta^{(v)} = -\frac{2g(1-(1-2g)\Delta_N)((1-2g)\Delta_N+\Delta_T)}{\left(1-(1-g)^2\right)\left(1+\frac{1}{1+g}(g\Delta_T-(1-2g)\Delta_N)\right)} \approx -2g((1-2g)\Delta_N+\Delta_T). \tag{2.37c}$$

In Eq. (2.37a), anisotropic parameters $\gamma^{(v)}$, $\varepsilon^{(v)}$ and $\delta^{(v)}$ are always negative in a HTI model, thus the fracture weakness parameters always have nonnegative values. The fracture characterization can be determined from the surface seismic signatures through the anisotropic representations in the seismic data.

2.3.8 Orthogonally Fractured Medium Moduli (VTI + HTI)

A reconstructed effective orthorhombic medium (the symmetry of a brick) is usually a combination of a VTI medium and a HTI medium that has been considered as a more realistic model to characterize the reservoir (Fig. 2.13). The reason is that geological sediments deposited into a horizontally layered medium show transversely isotropic elastic behavior with a vertical symmetry axis (VTI), while the maximum principal stress is a vertical one of compression from the overburden deposition that generated the vertical fractures with a horizontal symmetry axis. Therefore, an orthorhombic medium model is a more reasonable model to describe the subsurface structures. Based on the Schoenberg and Muir calculus, an orthogonally fractured medium can be treated as a vertical slip interface plus a VTI host medium

$$\boldsymbol{G}_{\mathbf{Orth}} = \boldsymbol{G}_{H\_\mathrm{VTI}} + \boldsymbol{G}_{v\_f}. \tag{2.38}$$

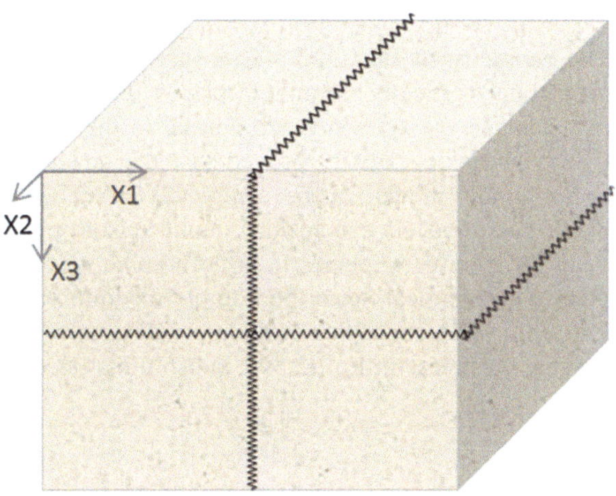

Fig. 2.13 Orthogonally fractured medium model. It is formed by a vertical fracture interface and horizontally fractured host medium

$\boldsymbol{G}_{\boldsymbol{H\_}\mathbf{VTI}}$ is the group stiffness of the VTI host medium that could include the stiffness of a horizontally fractured medium in Eq. (2.36) as well. $\boldsymbol{G}_{\boldsymbol{v\_f}}$ is the group element of the vertical fracture and has the same expression as in Eq. (2.34) (Appendix A).

$$
\boldsymbol{C}_{\mathbf{Orth}} = \begin{bmatrix} C_{11H\_\mathrm{VTI}}(1-\Delta_{\mathrm{N}}) & C_{12H\_\mathrm{VTI}}(1-\Delta_{\mathrm{N}}) & C_{13H\_\mathrm{VTI}}(1-\Delta_{\mathrm{N}}) & 0 & 0 & 0 \\ C_{12H\_\mathrm{VTI}}(1-\Delta_{\mathrm{N}}) & C_{11H_{\mathrm{VTI}}}\left(1-\frac{C_{12H\_\mathrm{VTI}}^2}{C_{11H\_\mathrm{VTI}}^2}\Delta_{\mathrm{N}}\right) & C_{13H\_\mathrm{VTI}}\left(1-\frac{C_{12H\_\mathrm{VTI}}}{C_{11H\_\mathrm{VTI}}}C_{12H\_\mathrm{VTI}}\Delta_{\mathrm{N}}\right) & 0 & 0 & 0 \\ C_{31H\_\mathrm{VTI}}(1-\Delta_{\mathrm{N}}) & C_{13H\_\mathrm{VTI}}\left(1-\frac{C_{112H\_\mathrm{VTII}}}{C_{11H\_\mathrm{VTI}}}\Delta_{\mathrm{N}}\right) & C_{33H\_\mathrm{VTI}}\left(1-\frac{C_{13H\_\mathrm{VTI}}^2}{C_{11H\_\mathrm{VTI}}C_{33H\_\mathrm{VTI}}}\Delta_{\mathrm{N}}\right) & 0 & 0 & 0 \\ 0 & 0 & 0 & C_{44H\_\mathrm{VTI}I} & 0 & 0 \\ 0 & 0 & 0 & 0 & C_{44H\_\mathrm{VTI}}(1-\Delta_V) & 0 \\ 0 & 0 & 0 & 0 & 0 & C_{66H\_\mathrm{VTI}}(1-\Delta_H) \end{bmatrix}. \tag{2.39}
$$

Equation (2.39) gives the stiffness of an orthogonally fractured medium. An effective orthogonally fractured medium has nine independent moduli that depend on the five independent moduli of the TI host medium and the three fracture compliances of the fracture. A more complex method for solving for the stiffness of the orthorhombic medium is given in Appendix A by considering a composite of a VTI medium and a HTI anisotropic media.

2.3.8.1 Orthorhombic Fractures Anisotropy

The geometry of Fig. 2.13 shows that the orthorhombic fracture model has two orthogonal vertically symmetric planes $[x_2, x_3]$, $[x_1, x_3]$ and one horizontally symmetric plane $[x_1, x_2]$ which are normal to the x_1-axis, x_2-axis and x_3-axis, respectively. Tsvankin (1997b) proved that the anisotropy in the symmetric planes of orthorhombic media can be completely described by known intrinsic anisotropy parameters, and all conclusions about anisotropy in VTI and HTI media remain valid for symmetric planes in orthorhombic media. If the properties of all vertical planes are identical, an orthorhombic medium can be reduced to a VTI medium. The stiffness tensor of this orthorhombic fractured medium is expressed through the fracture weakness in Eq. (2.39). Substituting necessary elements in Eq. (2.39) into

Eq. (2.17a) yields the anisotropic coefficients for the vertical symmetry $[x_2, x_3]$-plane

$$\gamma^{(1)} = \gamma_H + \frac{\Delta_V - \Delta_H}{2}, \tag{2.40a}$$

$$\varepsilon^{(1)} = \varepsilon_H, \tag{2.40b}$$

$$\delta^{(1)} = \delta_H. \tag{2.40c}$$

The subscript H of anisotropic parameters denotes the host medium, a horizontally fractured medium. Therefore, the anisotropic coefficients in the $[x_2, x_3]$-plane coincide with those of the horizontally fractured VTI host medium. It is seen that the $[x_2, x_3]$-plane represents an isotropic plane with a horizontal symmetric x_1-axis in this orthorhombic medium in which all of the velocities are not affected by the fracture and remain constant for all propagation directions (Tsvankin 1997b). If the fracture system is rotationally invariant, we have $\Delta_V = \Delta_H$, and then $\gamma^{(1)} = \gamma_H$.

Substituting some elements in Eq. (2.39) into Eq. (2.18a) gives the anisotropic coefficients for another vertical symmetry $[x_1, x_3]$-plane that is normal to the x_2-axis

$$\gamma^{(2)} = \gamma_H - \frac{\Delta_H}{2}, \tag{2.41a}$$

$$\varepsilon^{(2)} = \varepsilon_H - 2g(1-g)\Delta_\mathrm{N}, \tag{2.41b}$$

$$\delta^{(2)} = \delta_H - 2g[(1-2g)\Delta_\mathrm{N} + \Delta_V]. \tag{2.41c}$$

Note that the $[x_1, x_3]$-plane involves two distinguishing anisotropic coefficients. One is the corresponding anisotropic coefficient of the horizontally fractured VTI host medium with subscript "*H*", while the other anisotropic coefficient that results from the vertical fracture that approximately equals the coefficients $\gamma^{(v)}, \varepsilon^{(v)}, \delta^{(v)}$ describing a vertically fractured HTI model. This composite result is what we expected.

The anisotropic coefficients of the horizontal symmetry $[x_1, x_2]$-plane are obtained by substituting necessary elements in Eq. (2.39) into Eq. (2.19)

$$\delta^{(3)} = 2g[\Delta_\mathrm{N} - \Delta_V]. \tag{2.42}$$

Note that $\delta^{(3)}$ does not relate to the anisotropic coefficients of host medium because the horizontal symmetry $[x_1, x_2]$-plane is an isotropic plane in the horizontally fractured VTI host media.

2.4 Boundary Conditions

Boundary condition stands for a constraint function that must be satisfied along a boundary (Sheriff 2002). At the boundary, all waves are constrained such that kinematic displacements and dynamic stresses on one side of the boundary are related to displacements and stresses on the other side by certain boundary conditions. The direction of the boundary and contact properties of the two sides of the boundary decide the wave expressions in terms of the displacement and stress. Table 2.2 provides boundary types in 3D and the corresponding expressions of the boundary conditions in terms of displacement $\boldsymbol{u}$ and the stresses $\boldsymbol{\sigma}$, in which $\boldsymbol{u}^{\pm} = u_i^{\pm}, \boldsymbol{\sigma} = \sigma_{ij}^{\pm}.\ i = j = x, y, z.$

Table 2.2 Boundary conditions in 3D

Type of the boundaries	Welded boundary conditions	Nonwelded boundary conditions
z-normal boundary 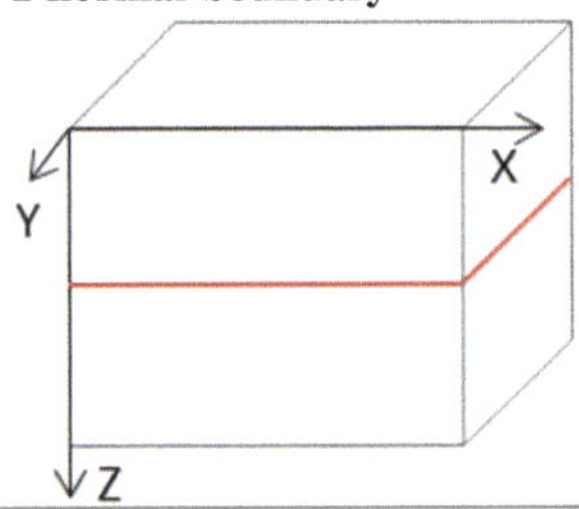	$\begin{cases} u_x^+ - u_x^- = 0 \\ \sigma_{xz}^+ - \sigma_{xz}^- = 0 \end{cases}$ $\begin{cases} u_z^+ - u_z^- = 0 \\ \sigma_{zz}^+ - \sigma_{zz}^- = 0 \end{cases}$ $\begin{cases} u_z^+ - u_z^- = 0 \\ \sigma_{zz}^+ - \sigma_{zz}^- = 0 \end{cases}$	$\begin{cases} u_x^+ - u_x^- = S_T\sigma_{xz}^+ \\ \sigma_{xz}^+ = \sigma_{xz}^- \end{cases}$ $\begin{cases} u_z^+ - u_z^- = S_N\sigma_{zz}^+ \\ \sigma_{zz}^+ = \sigma_{zz}^- \end{cases}$ $\begin{cases} u_y^+ - u_y^- = S_T\sigma_{yz}^+ \\ \sigma_{yz}^+ = \sigma_{yz}^- \end{cases}$
x-normal boundary 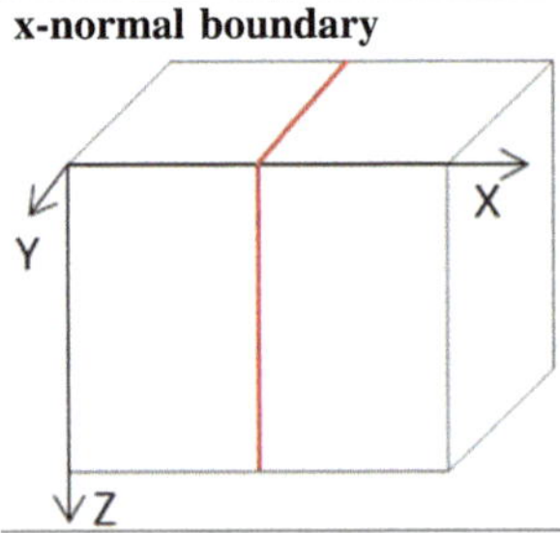	$\begin{cases} u_x^+ - u_x^- = 0 \\ \sigma_{xx}^+ - \sigma_{xx}^- = 0 \end{cases}$ $\begin{cases} u_z^+ - u_z^- = 0 \\ \sigma_{zx}^+ - \sigma_{zx}^- = 0 \end{cases}$ $\begin{cases} u_y^+ - u_y^- = 0 \\ \sigma_{yx}^+ - \sigma_{yx}^- = 0 \end{cases}$	$\begin{cases} u_x^+ - u_x^- = S_N\sigma_{xx}^+ \\ \sigma_{xx}^+ = \sigma_{xx}^- \end{cases}$ $\begin{cases} u_z^+ - u_z^- = S_T\sigma_{zx}^+ \\ \sigma_{zx}^+ = \sigma_{zx}^- \end{cases}$ $\begin{cases} u_y^+ - u_y^- = S_T\sigma_{yx}^+ \\ \sigma_{yx}^+ = \sigma_{yx}^- \end{cases}$
y-normal boundary 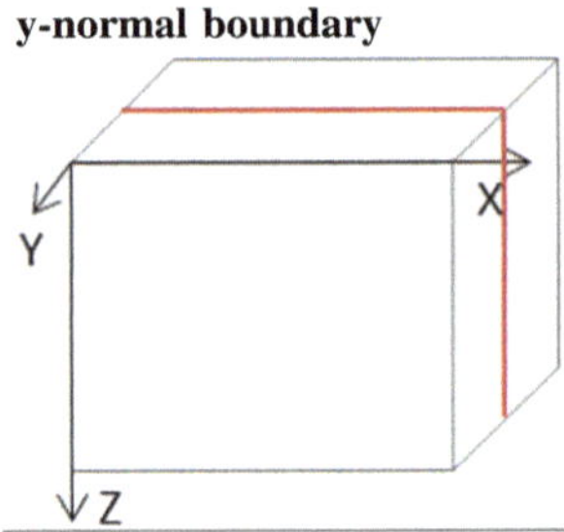	$\begin{cases} u_x^+ - u_x^- = 0 \\ \sigma_{xy}^+ - \sigma_{xy}^- = 0 \end{cases}$ $\begin{cases} u_z^+ - u_z^- = 0 \\ \sigma_{zy}^+ - \sigma_{zy}^- = 0 \end{cases}$ $\begin{cases} u_y^+ - u_y^- = 0 \\ \sigma_{yy}^+ - \sigma_{yy}^- = 0 \end{cases}$	$\begin{cases} u_x^+ - u_x^- = S_T\sigma_{xy}^+ \\ \sigma_{xy}^+ = \sigma_{xy}^- \end{cases}$ $\begin{cases} u_z^+ - u_z^- = S_T\sigma_{zy}^+ \\ \sigma_{zy}^+ = \sigma_{zy}^- \end{cases}$ $\begin{cases} u_y^+ - u_y^- = S_N\sigma_{yy}^+ \\ \sigma_{yy}^+ = \sigma_{yy}^- \end{cases}$

2.4.1 *Perfectly Welded Contact Interface*

Consider a compressional wave traveling in the x–z plane. It impinges upon a perfectly welded contact interface at which the kinematic displacements and the dynamic stresses of the wave quantities on the two sides of the interface satisfy

$$\boldsymbol{u}^{+} = \boldsymbol{u}^{-}, \tag{2.43}$$

$$\boldsymbol{\sigma}^{+} = \boldsymbol{\sigma}^{-}, \tag{2.44}$$

where $\boldsymbol{u}$ is kinematic displacement and $\boldsymbol{\sigma}$ is dynamic stress. $\boldsymbol{u}^{+} = \begin{Bmatrix} u_x^{+} \\ u_z^{+} \end{Bmatrix}, \boldsymbol{u}^{-} = \begin{Bmatrix} u_x^{-} \\ u_z^{-} \end{Bmatrix}, \boldsymbol{\sigma}^{+} = \begin{Bmatrix} \sigma_{xz}^{+} \\ \sigma_{zz}^{+} \end{Bmatrix}, \boldsymbol{\sigma}^{-} = \begin{Bmatrix} \sigma_{xz}^{-} \\ \sigma_{zz}^{-} \end{Bmatrix}$. The superscripts "–" and "+" denote the two sides of the interface (Fig. 2.14). The subscripts "x and z" or "xz and zz" denote the $\boldsymbol{u}$ or $\boldsymbol{\sigma}$ in tangential and normal components. For example, u_x^{+} or u_z^{+} can be interpreted as a displacement $\boldsymbol{u}$ in tangential x or normal z component at the "+" side of the interface. Equations (2.43) and (2.44) must be satisfied by the incident, reflected, and transmitted waves at the interface, and are known as the perfect welded contact boundary conditions. They state that both kinematic displacements and dynamic stresses of the wave are continuous across the perfect welded interface when the wave propagates at the interface. Figure 2.14 shows the perfectly welded interface boundary conditions.

2.4.1.1 Reflections and Transmissions

In seismology, an incident compressional plane wave produces reflected and transmitted waves at the interface. Figure 2.15 shows all elastic waves at the

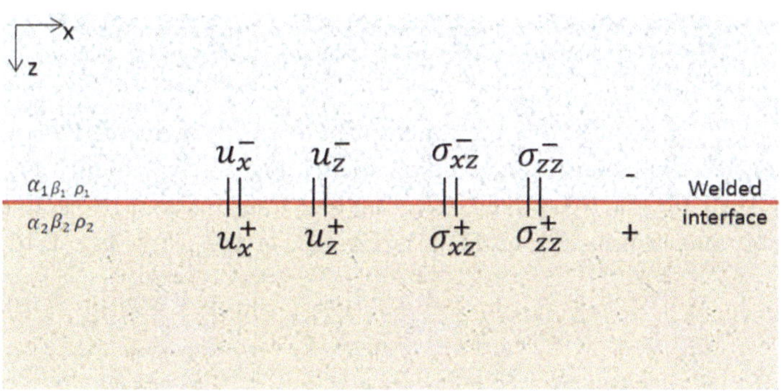

Fig. 2.14 Perfectly welded interface boundary conditions. $\boldsymbol{u}^{+} = \boldsymbol{u}^{-}, \boldsymbol{\sigma}^{+} = \boldsymbol{\sigma}^{-}$. The kinematic displacements and dynamic stresses are continuous across the welded interface

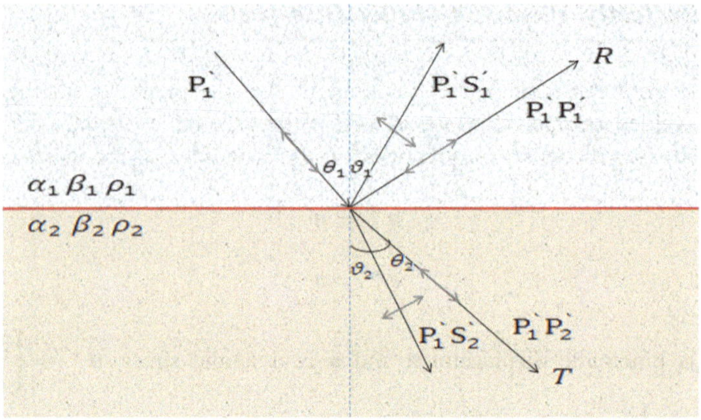

Fig. 2.15 Incident P-wave, reflected and transmitted PP and PS waves at a perfectly welded contact interface. $\boldsymbol{\theta}_1, \boldsymbol{\theta}_2, \vartheta_1$ and ϑ_2 are PP and PS-waves reflection and transmission angles respectively. The *single arrows* point in the direction of wave propagation. The *double arrows* indicate the direction of the wave polarization. $\boldsymbol{\alpha}_n, \boldsymbol{\beta}_n$ and $\boldsymbol{\rho}_n$, $\boldsymbol{n} = 1, 2$ are the media parameters

reflector in a homogeneous isotropic medium. The upper prime “‘” indicates up going reflection waves, and the down prime “‘” means down going waves that include the incident wave in the upper medium and the transmitted waves in the lower medium. Consider a plane wave,

$$\boldsymbol{u} = Ae^{-i\omega(t-\boldsymbol{s}\cdot\boldsymbol{x})}\boldsymbol{d}, \tag{2.45}$$

where A is the amplitude and is assumed to be unity, and $\boldsymbol{s}\cdot\boldsymbol{x} = s_x x + s_y y + s_z z$. Equation (2.45) represents a harmonic plane wave traveling in the $\boldsymbol{s}$ direction, where $\boldsymbol{s}$ is the slowness. $\boldsymbol{d}$ denotes the wave polarization, with $|\boldsymbol{d}| = 1$.

Substituting Eq. (2.45) into Eq. (2.43) and using Fig. 2.15, gives

$$\mathrm{P}_1^{\backprime}\sin\theta_1 = -\mathrm{P}_1^{\backprime}\sin\theta_1 - \mathrm{S}_1^{\prime}\cos\vartheta_1 + \mathrm{P}_2^{\backprime}\sin\theta_2 + \mathrm{S}_2^{\backprime}\cos\vartheta_2. \tag{2.46a}$$

$$\mathrm{P}_1^{\backprime}\cos\theta_1 = \mathrm{P}_1^{\prime}\cos\theta_1 - \mathrm{S}_1^{\prime}\sin\vartheta_1 + \mathrm{P}_2^{\backprime}\cos\theta_2 - \mathrm{S}_2^{\backprime}\sin\vartheta_2. \tag{2.46b}$$

where $\mathrm{P}_1^{\backprime}, \mathrm{P}_1^{\backprime}, \mathrm{P}_2^{\backprime}, \mathrm{S}_1^{\backprime}$ and $\mathrm{S}_2^{\backprime}$ are the amplitudes of the P-incident, PP-reflected and transmitted, PS-reflected and transmitted waves, respectively, while $\theta_1, \theta_2, \vartheta_1$ and ϑ_2 are angle of PP and PS-waves reflection and transmission, respectively.

Similarly, substituting Eq. (2.45) into Eq. (2.44) and using Fig. 2.15 gives

$$\mathrm{P}_1^{\backprime}x_1\cos\theta_1 = \mathrm{P}_1^{\prime}x_1\cos\theta_1 + \mathrm{S}_1^{\prime}\beta_1 r_1 + \mathrm{P}_2^{\backprime}x_2\cos\theta_2 + \mathrm{S}_2^{\backprime}\beta_2 r_2. \tag{2.47a}$$

$$\mathrm{P}_1^{\backprime}\alpha_1 r_1 = -\mathrm{P}_1^{\prime}\alpha_1 r_1 + \mathrm{S}_1^{\prime}x_1\cos\vartheta_1 + \mathrm{P}_2^{\backprime}\alpha_2 r_2 - \mathrm{S}_2^{\backprime}x_2\cos\vartheta_2. \tag{2.47b}$$

where α_n, β_n and ρ_n are the P-wave velocity, S-wave velocity and density, respectively. $x_n = 2\rho_n\beta_n^2\mathrm{P}$, $r_n = \rho_n\left(1 - 2\beta_n^2\mathrm{P}^2\right), n = 1, 2$ (Aki and Richards 1980). P is the ray parameter.

$$\frac{\sin\theta_n}{\alpha_n} = \frac{\sin\vartheta_n}{\beta_n} = \mathrm{P}. \tag{2.48}$$

2.4.1.2 Zoeppritz Equations

Equations (2.46a) and (2.47a) are four linear equations in four unknowns, and they can be used to solve for the four unknown PP and PS reflection transmission coefficients in a homogeneous isotopic media. This system of equations can be written in matrix form as

$$\begin{bmatrix} \alpha_1\mathrm{P} \\ \cos\theta_1 \\ x_1\cos\theta_1 \\ \alpha_1 r_1 \end{bmatrix} = \begin{bmatrix} -\alpha_1\mathrm{P} & -\cos\vartheta_1 & \alpha_2\mathrm{P} & \cos\vartheta_2 \\ \cos\theta_1 & -\beta_1\mathrm{P} & \cos\theta_2 & -\beta_2\mathrm{P} \\ x_1\cos\theta_1 & \beta_1 r_1 & x_2\cos\theta_2 & \beta_2 r_2 \\ -\alpha_1 r_1 & x_1\cos\vartheta_1 & \alpha_2 r_2 & -x_2\cos\vartheta_2 \end{bmatrix} \begin{bmatrix} \mathrm{P}'\mathrm{P}_1' \\ \mathrm{P}'\mathrm{S}_1' \\ \mathrm{P}'\mathrm{P}_2' \\ \mathrm{P}'\mathrm{S}_2' \end{bmatrix}. \tag{2.49}$$

These equations are known as the Zoeppritz equations (1919). The equations reveal that the amplitudes of reflected and transmitted waves are functions of the angles at the interface. For a normal incidence case, the PP-wave reflection and transmission coefficients are

$$R_w = \frac{\mathrm{P}_1'}{\mathrm{P}_1'} = \mathrm{P}'\mathrm{P}_1' = \frac{\alpha_2\rho_2 - \alpha_1\rho_1}{\alpha_2\rho_2 + \alpha_1\rho_1}, \tag{2.50a}$$

$$T_w = \frac{\mathrm{P}_2'}{\mathrm{P}_1'} = \mathrm{P}'\mathrm{P}_2' = \frac{2\alpha_1\rho_1}{\alpha_2\rho_2 + \alpha_1\rho_1}. \tag{2.50b}$$

Note that there should be only transmitted waves passing through the interface in a uniform homogeneous isotropic medium at normal incidence since in that case $\alpha_2\rho_2 - \alpha_1\rho_1 = 0$.

2.4.2 *Imperfectly Welded (Nonwelded) Contact Interface*

Consider a compressional wave hitting upon a deformed (fractured) linear slip interface that is different from the perfect welded interface. At this slip interface, the incident, reflected, and transmitted waves are constrained by the imperfectly welded (nonwelded) contact boundary conditions: the dynamic stresses are continuous across the interface, but the kinematic displacements are discontinuous and the

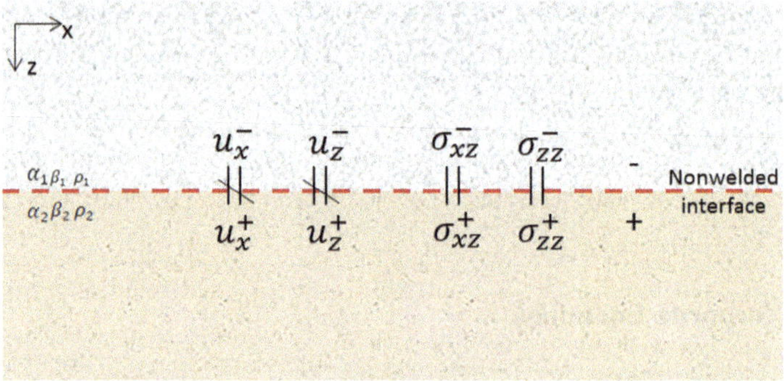

Fig. 2.16 Imperfectly welded (nonwelded) contact interface boundary conditions. $\boldsymbol{u}^+ \neq \boldsymbol{u}^-, \boldsymbol{\sigma}^+ = \boldsymbol{\sigma}^-$ The kinematic displacements are discontinuous, but the dynamic stresses are continuous across the nonwelded contact interface

differential displacements are linearly proportional to the corresponding stresses (Fig. 2.16):

$$\boldsymbol{u}^+ - \boldsymbol{u}^- = \boldsymbol{S}\boldsymbol{\sigma}^+ \tag{2.51}$$

$$\boldsymbol{\sigma}^+ = \boldsymbol{\sigma}^- \tag{2.52}$$

where $\boldsymbol{u}$ and $\boldsymbol{\sigma}$ are the same as in Eqs. (2.43, 2.44). The parameters $\boldsymbol{S} = \begin{cases} S_\mathrm{T} \\ S_\mathrm{N} \end{cases}$, S_T and S_N have been described in Eq. (2.27) as the tangential and normal fracture compliances of the fractured medium (Schoenberg 1980), which are the reciprocals of the rock stiffnesses. S_T is for a normally incident shear wave, and S_N is for a normally incident compressional wave (Schoenberg 1980). S_T and S_N imply that fracture deformation is a combination of tangential deformation and normal deformation.

As Eq. (2.52) shows, the stress is continuous across the interface, even though it is a nonwelded contact interface. If it was not, the equation of motion,

$$\frac{\partial \sigma_{ij}}{\partial x_j} = \rho \frac{\partial^2 u_i}{\partial t^2}, \tag{2.52a}$$

indicates that forces and accelerations would be infinite for a discontinuity in stress, which is unphysical.

2.4.2.1 Reflections and Transmissions

Similarly, a plane compressional incident wave produces reflected and transmitted waves at the linear slip interface in a homogeneous isotropic medium (Fig. 2.16).

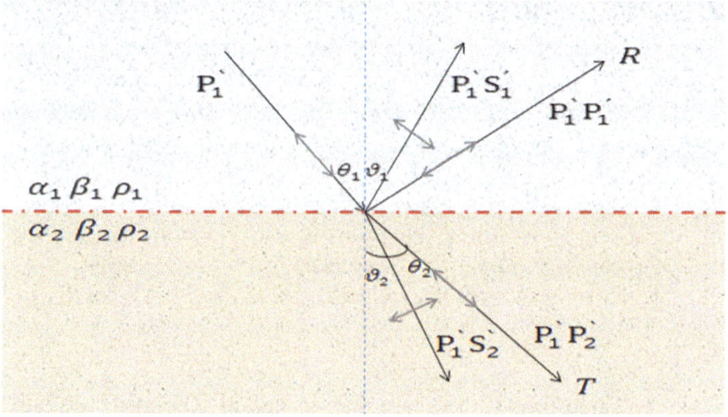

Fig. 2.17 Incident P-wave, reflected and transmitted PP and PS waves at the imperfectly welded (nonwelded) contact interface. $\boldsymbol{\theta}_1, \boldsymbol{\theta}_2, \vartheta_1$ and ϑ_2 are PP and PS-waves reflection and transmission angles respectively. The *single arrows* point in the direction of wave propagation. The *double arrows* indicate the direction of the wave polarization. $\boldsymbol{\alpha}_n, \boldsymbol{\beta}_n$ and $\boldsymbol{\rho}_n$. $\boldsymbol{n} = 1, 2$ are the media parameters. $\boldsymbol{S}_\mathrm{T}$ and $\boldsymbol{S}_\mathrm{N}$ are tangential and normal fracture compliances, respectively

For a given frequency ω, substituting Eq. (2.45) into Eq. (2.51)and using Fig. 2.17 gives

$$\begin{aligned} P_1^{`} \sin\theta_1 = & - P_1' \sin\theta_1 - \mathrm{S}_1' \cos\vartheta_1 + P_2^{`}(\sin\theta_2 - i\omega S_\mathrm{T} x_2 \cos\theta_2) \\ & + \mathrm{S}_2^{`}(\cos\vartheta_2 - i\omega S_\mathrm{T} \beta_2 r_2), \end{aligned} \tag{2.53a}$$

$$\begin{aligned} \mathrm{P}_1^{`} \cos\theta_1 = & \, \mathrm{P}_1' \cos\theta_1 - \mathrm{S}_1' \sin\vartheta_1 + \mathrm{P}_2^{`}(\cos\theta_2 - i\omega S_\mathrm{N} \alpha_2 r_2) \\ & + \mathrm{S}_2^{`}(-\sin\vartheta_2 + i\omega S_\mathrm{N} x_2 \cos\vartheta_2). \end{aligned} \tag{2.53b}$$

Similarly, substituting the plane wave in Eq. (2.45) into Eq. (2.52) for the stress nonwelded boundary conditions, gives

$$\mathrm{P}_1^{`} x \cos\theta_1 = \mathrm{P}_1' x_1 \cos\theta_1 + \mathrm{S}_1' \beta_1 r_1 + \mathrm{P}_2^{`} x_2 \cos\theta_2 + \mathrm{S}_2^{`} \beta_2 r_2, \tag{2.54a}$$

$$\mathrm{P}_1^{`} \alpha_1 r_1 = -\mathrm{P}_1' \alpha_1 r_1 + \mathrm{S}_1' x_1 \cos\vartheta_1 + \mathrm{P}_2^{`} \alpha_2 r_2 - \mathrm{S}_2^{`} x_2 \cos\vartheta_2, \tag{2.54b}$$

where, the symbols $\mathrm{P}_1^{`}, \mathrm{P}_1', \mathrm{P}_2^{`}, \mathrm{S}_1', \mathrm{S}_2^{`}$ and $\alpha_n, \beta_n, \rho_n$ and $x_n = 2\rho_n \beta_n^2 \mathrm{P}, r_n = \rho_n(1 - 2\beta_n^2 \mathrm{P}^2), n = 1, 2$ in Eqs. (2.46a) and (2.47a) are the same as in Eqs. (2.53a) and (2.54a). Note that Eq. (2.54a) are identical to Eq. (2.47a). An explanation for this is that the dynamic stresses always obey the law of conservation of energy, regardless of whether contact of the interface is perfectly welded or not.

2.4.2.2 Zoeppritz Equations

Combining Eqs. (2.53a) and (2.54a) yields a system of four linear equations in four unknowns. They can be used to solve for the four unknown reflection and transmission coefficients of the PP and PS-waves at the imperfect welded interface:

$$\begin{bmatrix} \alpha_1 P \\ \cos\theta_1 \\ x_1\cos\theta_1 \\ \alpha_1 r_1 \end{bmatrix} = \begin{bmatrix} -\alpha_1 \mathrm{P} & -\cos\vartheta_1 & \alpha_2\mathrm{P} - i\omega S_T x_2\cos\theta_2 & \cos\vartheta_2 - i\omega S_T\beta_2 r_2 \\ \cos\theta_1 & -\beta_1\mathrm{P} & \cos\theta_2 - i\omega S_N\alpha_2 r_2 & -\beta_2\mathrm{P} + i\omega S_N x_2\cos\vartheta_2 \\ x_1\cos\theta_1 & \beta_1 r_1 & x_2\cos\theta_2 & \beta_2 r_2 \\ -\alpha_1 r_1 & x_1\cos\vartheta_1 & \alpha_2 r_2 & -x_2\cos\vartheta_2 \end{bmatrix} \begin{bmatrix} \mathrm{P}_1^{`}\mathrm{P}_1' \\ \mathrm{P}_1^{`}\mathrm{S}_1' \\ \mathrm{P}_1^{`}\mathrm{P}_2^{`} \\ \mathrm{P}_1^{`}\mathrm{S}_2^{`} \end{bmatrix} \quad (2.55)$$

where $i = \sqrt{-1}$. Equations (2.55), are the so-called Zoeppritz equations for the linear slip interface, and these equations are important for fracture analysis because they reveal the relationship between the wave incident angles and the reflection/transmission coefficients at the linear slip interface. Note that they have the same pattern as the original Zoeppritz equations (1919). Equation (2.55) also shows that the reflection/transmission coefficients are frequency dependent with the fracture parameters S_T and S_N (Schoenberg 1980; Chaisri and Krebes 2000; Chaisri 2002). Pyrak-Nolte et al. (1990a, b) and Chaisri and Krebes (2000) have shown that the reflected P-wave has a lower amplitude at the lower frequencies that has an apparent attenuation in the nonwelded interface.

Chaisri and Krebes (2000) and Chaisri (2002) give the conditions that the fracture parameters and the frequency should satisfy the following conditions.

$$\omega S_N \rho_2 \alpha_2 < < 1, \quad (2.56a)$$

and

$$\omega S_T \rho_2 \beta_2 < < 1. \quad (2.56b)$$

From Eq. (2.55), the normal incidence P-wave reflection and transmission coefficients at the slip interface in a homogeneous isotropic medium are

$$R_{\mathrm{non\_}w} = \mathrm{P}_1^{`}\mathrm{P}_1' = \frac{Z_2 - Z_1 + i\omega\, S_\mathrm{N} Z_2 Z_1}{Z_2 + Z_1 - i\omega\, S_\mathrm{N} Z_2 Z_1}, \quad (2.57a)$$

$$T_{\mathrm{non\_}w} = \mathrm{P}_1^{`}\mathrm{P}_2^{`} = \frac{2Z_1}{Z_2 + Z_1 - i\omega\, S_\mathrm{N} Z_2 Z_1}, \quad (2.57b)$$

where Z_n is the media impedance: $Z_n = \rho_n \alpha_n$. Note that reflection coefficients at a linear slip interface should be nonzero at normal incidence even if the interface is in the uniform homogeneous isotropic medium where there is no impedance contrast, i.e., $Z_2 - Z_1 = 0$. This is a substantial difference between a linear slip interface and a welded interface. In other words, reflections from the slip interface show that a fracture itself generates a reflection which is due to the displacement discontinuity, rather than an impedance contrast at the boundary.

2.5 Conclusions

This chapter theoretically and systematically provides a link between the inherent elastic properties of the medium and the gained fracture compliance parameters of the fractured medium. Also it presents the knowledge of fracture formation, fracture parameters, and fracture detection. As well, it supplies a rock stiffness composition and decomposition framework within which a complex fractured medium stiffness can be simplified into a fracture stiffness and a host medium stiffness, and vice versa. This infers that a medium forming a fracture feature combined with a host medium is equivalent to a fractured medium, and vice versa. Additionally, it discusses the boundary conditions that are constraints for all waves at an interface, thus it is a key for next chapter's study, which is on the seismic signature response of the fracture.

Elastic tensors are the basis of understanding what will happen to a rock when it is subjected to force. In general, an elastic rock deformation first obeys Hooke's law as a reversible deformation when it is subjected to a small force, then it changes to an irreversible ductile deformation if the force keeps exerting. Once the rock strain exceeds a strain threshold, the rock deformation finally changes to a physically permanent fracture deformation. A discussion of stress and strain is presented in Sect. 2.1 for illustrating the different deformation stages. As a subsurface medium, the fracture formation similarly suffers the same deformation stages as a rock because the medium still undergoes the geological movement and overburden compressions. Commonly, most geological fractures are vertical or nearly vertical fractures because the overburden compression is the maximum stress rather than the other two horizontal stresses.

From a geophysical viewpoint, the fracture orientation and density are significantly more important parameters than the fracture openings in reservoir characterizations. The orientation is determined by the directions of maximum and minimum stress. The direction of the fracture is parallel to the direction of the maximum compressive stress and perpendicular to the direction of the minimum compressive stress. The fracture density is a measurement of the number of the fractures per meter along a certain direction, and it influences permeability evaluations: the higher the fracture density, the higher the permeability if the fractures are conductive.

The fracture detection usually is an integrated method involving geology, geophysics, and petrophysics technologies. For this integrated method, petrophysics data are regarded as the most reliable, i.e., as "hard data" to be broadly applied since the data are directly observed with higher resolution from the reservoir, even though the coverage is sparser that seismic data coverage. Petrophysics data such as core analysis, temperature log, caliper log, density log, dipmeter log, and image log data have been individually introduced in Sect. 2.2.3 in order to have a deep knowledge of well logs that predict the fractures in the reservoir.

Backus (1962) average theory is the basis of forming effective media. A perfectly welded layered medium can be approximated as an effective TI medium (a transversely isotropic medium with a symmetric axis). It indicates that an algebraic calculation is probably applicable for the rock stiffnesses of a medium. In Sect. 2.3, three effective fractured anisotropic media stiffnesses (VTI, HTI, and orthorhombic) have been calculated and presented. The calculations show that a fractured rock stiffness can be algebraically decomposed into the fracture stiffness and the host medium stiffness, and vice versa. Analogously, a fractured medium is equivalent to a fracture feature combined with a homogeneous host medium, whereas a fracture feature plus a homogeneous host medium is equivalent to a fractured medium. In other words, a fracture can be obtained by taking a fractured medium and subtracting the host medium, or, a host medium can be obtained by getting rid of the fracture from the fractured medium. In some sense, the fractured medium with five independent moduli exhibits properties of a transversely isotropic medium with a symmetry axis.

The linear slip interface theory (Schoenberg 1980) has been studied and it is regarded as one of perfect fracture model. Pyrak-Nolte et al. (1990a, b) and Hsu and Schoenberg (1993) conducted laboratory experiments to validate the model. As a result, a horizontally fractured medium is effectively formed by a linear slip interface embedded in a homogeneous isotropic host medium. A vertically fractured medium is effectively constructed by a vertical linear slip interface inserted in a homogeneous isotropic host medium. Furthermore, the orthogonally fractured media are effectively made from a vertical linear slip interface and a transversely isotropic medium with a vertical symmetric axis (VTI) and a homogeneous host medium.

The anisotropy problem has been discussed. Based on effective medium theory, the Schoenberg-Muir calculus theory and the mechanism of the fracture formation, it has been confirmed that fractured media have the characteristics of a transversely isotropic medium and that the anisotropy is an induced anisotropy caused by tectonic movement, rather than an intrinsic anisotropy formed by natural deposition.

The anisotropy issue has been discussed and it can be described by the parameters γ, ε and δ with five independent moduli (Thomsen 1986). The fractured medium has five independent moduli that can be expressed in terms of fracture weaknesses Δ_N and Δ_T that express the fracture in terms of induced anisotropic problem. There is a way to transform the intrinsic anisotropic parameters $\gamma, \varepsilon, \delta$ and fracture anisotropic description Δ_N and Δ_T into each other.

The boundary conditions control the seismic reflection and transmission coefficient values at the interface. Two sets of PP and PS reflection and transmission coefficients formulae have been presented in the last section. One is for a perfectly welded contact interface boundary and the other is for an imperfectly welded (nonwelded) contact interface boundary. The imperfectly welded (nonwelded) boundary conditions require that the kinematic displacements are discontinuous across the interface, whereas the dynamic stresses are continuous across the interface. These boundary conditions completely constrain all waves at the fracture interface. It is also demonstrated that the reflection coefficients of the nonwelded boundary are frequency dependent. In addition, for a wave normally incident propagating in a uniform homogeneous isotropic medium, there is no reflection at a perfectly welded boundary that has no impedance contrast. However, there are reflections at an imperfectly welded (nonwelded) contact boundary, since a fracture manifests itself a reflector, and the reflected waves are caused by a displacement discontinuity across the fracture, instead of by the impedance contrast at the fracture interface.

This chapter mainly concludes that a linear slip interface can model a fracture regardless of its fracture shape or microstructure, and it satisfies the nonwelded contact boundary conditions when a wave impinges on it. The fractured medium is composed by the fracture and homogeneous isotropic host medium, and vice versa. The composed horizontally and vertically fractured medium with five independent moduli and the orthogonally fractured medium with nine independent moduli shows that the fractured medium present the medium properties of a transversely isotropic medium with a symmetric axis (TI).

References

Aguilera, R. (1998). Geologic aspects of naturally fractured reservoirs. *The Leading Edge, 17,* 1667–1670.

Aki, K., & Richards, P. G. (1980). *Quantitative seismology: Theory and methods* (Vol. 1). W.H. Freeman and Company.

Anderson, E. M. (1905). The dynamics of faulting. *Transactions of the Edinburgh Geological Society, 8,* 387–402.

Anderson, E. M. (1942/1951). *The dynamics of faulting and dyke formation with application to Britain*. Edinburgh: Oliver & Boyd.

Backus, G. E. (1962). Long-wave elastic anisotropy produced by horizontal layering. *Journal Geophysical Research, 67,* 4427–4440.

Bakulin, A., Grechka, V., & Tsvankin, I. (2000a). Estimation of fracture parameters from reflection seismic data-Part I: HTI model due to a single fracture set. *Geophysics, 65*(6), 1788–1802.

Bakulin, A., Grechka, V., & Tsvankin, I. (2000b). Estimation of fracture parameters from reflection seismic data-Part II: Fractured models with orthorhombic symmetry. *Geophysics, 65* (6), 1803–1817.

Bakulin, A., Grechka, V., & Tsvankin, I. (2000c). Estimation of fracture parameters from reflection seismic data-Part III: Fractured models with monoclinic symmetry. *Geophysics, 65*(6), 1818–1830.

Carcione, J. M., Kosloff, D., & Behle, A. (1991). Long wave anisotropy in stratified media: A numerical test. *Geophysics, 56,* 245–254.

Carcione, J. M., Picotti, S., Cavallini, F., & Santos, J. E. (2012). Numerical test of the Schoenberg-Muir theory. *Geophysics, 77,* C27–C35.

Chaisri, S. (2002). *A study of a nonwelded contact interface: Exact and approximate formulas for P-SV reflection and transmission coefficients, and frequency domain raytracing*. Ph.D. thesis, The University of Calgary.

Chaisri, S., & Krebes, E. S. (2000). Exact and approximate formulas for P-SV reflection and transmission coefficients for a nonwelded contact interface. *Journal Geophysical Research, 105,* 28045–28054.

Crampin, S. (1985). Evidence for aligned cracks in the Earth's crust. *First Break, 3*(3), 12–15.

Crampin, S., & Bamford, D. (1977). Inversion of P-wave velocity anisotropy. *Geophysical Journal Roy Astronomical Society, 49,* 123.

Cui, X., & Lines, L. R. (2011). PP PS reflection and transmission coefficients for a non-welded interface contact with anisotropic media. 2011 SEG annual meeting.

Hood, J. (1991). A simple method for decomposing fracture-induced anisotropy. *Geophysics, 56,* 1275–1279.

Hsu, C.-J., & Schoenberg, M. (1993). Elastic waves through a simulated fracture medium. *Geophysics, 58,* 964–977.

Krebes, E. S. (2006). *Seismic theory and methods.* Course Notes, The University of Calgary.

Lefeuvre, F. (1993). Fracture related anisotropy detection and analysis; And if the P-waves were enough. *SEG Expanded Abstracts, 64,* 942–945.

Lynn, H., Simon, K. M., Bates, R., & Van Dok, R. (1996). Azimuthal anisotropy in P-wave 3-D (multiazimuth) data. *Leading Edge, 15*(8), 923–928.

Nelson, S. A. (2003). *EENS 111, geophysical geology deformation of rock.* Tulane University. http://www.tulane.edu/~sanelson/geol111/deform.htm

Nichols, D., Muir, F., & Schoenberg, M. (1989). Elastic properties of rocks with multiple fracture sets. In *59th Annual International Meeting Society of Exploration Geophysicists, Expanded Abstracts* (pp. 471–474).

Peterson, J. E., Hopkins, D., & Myer, L. (1993). Application of seismic displacement discontinuity theory to Hanford data. In *63rd Annual International Meeting Society Exploration Geophysicists, Expanded Abstracts* (pp. 1006–1009).

Pyrak-Nolte, L. J., Myer, L. R., & Cook, N. G. W. (1990). Transmission of seismic waves across single natural fractures. *Journal of Geophysical Research, 95,* 8617–8638.

Pyrak-Nolte, L. J., Myer, L. R., & Cook, N. G. W. (1990b). Anisotropy in seismic velocities and amplitudes from multiple parallel fractures. *Journal Geophysical Research, 95*(B7), 11345–11358.

Ramos, A. C. B., & Davis, T. (1997). 3-D AVO analysis and modeling applied to fracture detection in coalbed methane reservoirs. *Geophysics, 62,* 1683–1695.

Rider, M. (2002). *The geological interpretation of well logs* (2nd ed.). Sutherland: Rider-French Consulting Ltd.

Rueger, A. (1996). Variation of P-wave reflectivity with offset and azimuth in anisotropic media. *SEG Annual Meeting Abstracts, 66,* 1810–1813.

Sayers, C. M., Taleghani, A. D., & Adachi, J. (2009). The effect of mineralization on the ratio of normal to tangential compliance of fractures. *Geophysical Prospecting, 57*(3), 439–446.

Schoenberg, M. (1980). Elastic wave behavior across linear slip interfaces. *Journal of the Acoustical Society of America, 68,* 1516–1521.

Schoenberg, M., & Muir, F. (1989). A calculus for finely layered anisotropic media. *Geophysics, 54,* 581–589.

Schoenberg, M., & Douma, J. (1988). Elastic wave propagation in media with parallel fractures and aligned cracks. *Geophysical Prospecting, 36,* 571–590.

Schoenberg, M., & Sayers, C. M. (1995). Seismic anisotropy of fractured rock. *Geophysics, 60,* 204–211.

Sheriff, R. E. (2002) *Encyclopedic dictionary of applied geophysics.* Geophysical Reference Series, 13. Tulsa, Oklahoma: Society of Exploration Geophysicists.

Singhal, B. B. S., & Gupta, R. P. (1999) *Applied hydrogeology of fractured rocks* (400 p.). Dordrecht, The Netherlands: Kluwer Academic Publishers.

Thomsen, L. (1986). Weak elastic anisotropy51. *Geophysics, 51*(10), 1954–1966.

Tsvankin, I. (1997a). Reflection moveout and parameter estimation for horizontal transverse isotropy. *Geophysics, 62,* 614–629.

Tsvankin, I. (1997b). Anisotropic parameters and P-wave velocity for orthorhombic media. *Geophysics, 62,* 1292–1309.

Winterstein, D. F. (1990). Velocity anisotropy terminology for geophysicists. *Geophysics, 55*(8), 1070–1088.

Zoeppritz, K. (1919). Erdbebenwellen VIIIB, On the reflection and propagation of seismic waves. *Göttinger Nachrichten*, I, 66.

Chapter 3
Seismic Forward Modeling of Fractures

Abstract Chapter 2 stated that the fractures play a very important role in the oil and gas production since they significantly affect the reservoir permeability, and the petrophysical detection technique is the most reliable in the integrated method for seeking fractures because it directly acquires data from the reservoir with a higher resolution. However, petrophysical data are extremely sparse and fracture patterns rapidly change, thus accurate delineation of fracture distributions in the subsurface imperatively demands geophysical seismic data that can provide global information coverage of the reservoir. The seismic measurements are wave reflection data that could imprint complex geological information with a relatively lower resolution. However, detecting fractures from seismic data is a big challenging problem because of the lack of understanding of the fracture responses in the intricate seismic data. Therefore, the main goals of this chapter are to address the problem by studying forward modeling of fractures to reveal the fracture representations in seismic data.

Keywords Finite-difference (FD) modeling · FD schemes · Fictitious points · Displacements–stresses · Perfectly welded boundary conditions · Imperfectly boundary conditions · Wave equation of motion · Horizontally fractured medium · Vertically fractured medium · Orthogonally fractured medium · Direction of the fracture · Amplitude · Traveltime · Fracture-induced anisotropy

3.1 Finite-Difference Forward Modeling Formulation Approaches

Seismic forward modeling usually attempts to simulate the rock properties of the earth model and the response of seismic wave propagation as the waves travel through the model (Lines and Newrick 2004). In seismology, the finite-difference (FD) method is widely used to numerically solve wave equation with a FD stencil to simulate seismic waves traveling through the structured subsurface. Kelly et al. (1976) presented a homogeneous formulation approach for solving the wave equation through FD modeling. In 1982, Korn and Stöckl (1982) used a

© Springer Nature Singapore Pte Ltd. 2018

X. Cui et al., *Seismic Forward Modeling of Fractures and Fractured Medium Inversion*, Springer Geophysics, DOI 10.1007/978-981-10-3584-5_3

homogeneous approach to model SH wave propagation in inhomogeneous media. An effective medium scheme (Coates and Schoenberg 1995) has been used to model the fracture with an explicit slip interface boundary condition. The explicit boundary condition approaches treat the fracture as a nonwelded contact interface, satisfying displacement discontinuity and stress continuity. With a homogeneous formulation approach and the explicit boundary conditions, Slawinski and Krebes (2002a, b) implemented the homogeneous finite-difference approach with additional fictitious points to model SH wave and P-SV wave propagation in the fractured media in which the fractures are aligned with the vertical or horizontal directions among the grids.

An advantage of the homogeneous approach with the explicit nonwelded boundary conditions is that it is able to constrain all the wave propagation in the homogeneous medium and satisfy the displacement discontinuity at the fracture boundary. Therefore, this approach is ideally selected to do the fractured medium forward modeling in which the boundary condition is imposed explicitly to govern wave propagation through the fractured medium.

3.2 Fictitious Grid and the Real Grid Points

In order to implement the fracture FD forward modeling with the homogeneous approach and explicitly impose the nonwelded boundary conditions, a fictitious displacement concept as an additional real displacement has been introduced for solving the homogeneous formulations (Korn and Stöckl 1982; Slawinski and Krebes 2002a, b). As shown in Fig. 3.1, the fictitious displacements are denoted by overhead tildes. The fictitious displacement points have physical grid locations corresponding to real displacement points such that the two media on either side of the interface extend into one another. The fictitious displacement $\tilde{\boldsymbol{u}}_{(x+\Delta x,z)}$ at grid point $(x+\Delta x, z)$, for instance, attempts to extend the medium at grid point (x, z) into the medium at grid point $(x+\Delta x, z)$ through the boundary at $X = (x+\frac{1}{2}\Delta x, z)$, while the fictitious displacement $\tilde{\boldsymbol{u}}_{(x,z)}$ at grid point (x, z) is the consequence of extending the medium at grid point $(x+\Delta x, z)$ into the medium at grid point (x, z) through the boundary $X = (x+\frac{1}{2}\Delta x, z)$. In Fig. 3.1, a real displacement $\boldsymbol{u}_{(x,z)}$ may relate to the four boundaries at $X = (x \pm \frac{1}{2}\Delta x, z)$ and $Z = (x, z \pm \frac{1}{2}\Delta z)$, the four nearest-neighbor real displacements $\boldsymbol{u}_{(x\pm\Delta x,z)}, \boldsymbol{u}_{(x,z\pm\Delta z)}$, the four next-nearest-neighbor real displacements $\boldsymbol{u}_{(x\pm\Delta x,z\pm\Delta z)}$ as well as fictitious displacements such as $\tilde{\boldsymbol{u}}_{(x\pm\Delta x,z)}, \tilde{\boldsymbol{u}}_{(x,z\pm\Delta z)}$ and $\tilde{\boldsymbol{u}}_{(x\pm\Delta x,z\pm\Delta z)}$. The bracket of the subscript x, z in symbol displacement are the grid indexes and Δx, Δz are the step size in the x-direction and z-direction, respectively. The uppercase letters Z and X are boundary indices for the z-normal boundary (the horizontal boundary that is normal to the z-axis) and the x-normal boundary (the vertical boundary that is normal to the x-axis), respectively. Here, $\boldsymbol{u} = \begin{cases} u_x \\ u_z \end{cases}$.

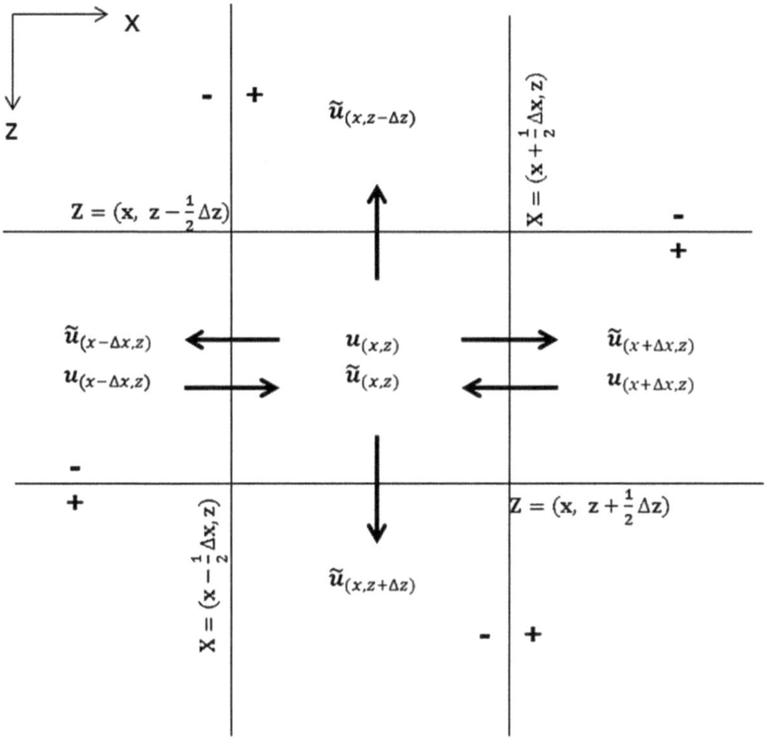

Fig. 3.1 Schematic of the fictitious displacements, the real displacement points, and the related boundaries. The fictitious displacements are denoted by *overhead tildes*. The signs "−" and "+" specify the side of the boundary respect to the x, z-axis. $\boldsymbol{u} = \begin{cases} u_x \\ u_z \end{cases}$

Korn and Stöckl (1982) assumed that a real grid displacement is approximately equal to the displacements at the sides of the boundary, i.e.,

$$\boldsymbol{u}^{+}_{(x,z-\frac{1}{2}\Delta z)} = \boldsymbol{u}_{(x,z)} = \boldsymbol{u}^{-}_{(x,z+\frac{1}{2}\Delta z)}, \tag{3.01a}$$

$$\boldsymbol{u}^{+}_{(x-\frac{1}{2}\Delta x,z)} = \boldsymbol{u}_{(x,z)} = \boldsymbol{u}^{-}_{(x+\frac{1}{2}\Delta x,z)}, \tag{3.01b}$$

and

$$\boldsymbol{u}_{(x,z+\Delta z)} = \boldsymbol{u}^{+}_{(x,z+\frac{1}{2}\Delta z)}, \tag{3.02a}$$

$$\boldsymbol{u}_{(x,z-\Delta z)} = \boldsymbol{u}^{-}_{(x,z-\frac{1}{2}\Delta z)}, \tag{3.02b}$$

$$\boldsymbol{u}_{(x+\Delta x,z)} = \boldsymbol{u}^{+}_{(x+\frac{1}{2}\Delta x,z)}, \tag{3.03a}$$

$$\boldsymbol{u}_{(x-\Delta x,z)} = \boldsymbol{u}^{-}_{(x-\frac{1}{2}\Delta x,z)}, \tag{3.03b}$$

where, the superscripts "−" and "+" in Eqs. (3.01a–3.03b) indicate two sides of the boundary with respect to the axis (Fig. 3.1).

3.3 Finite-Difference Scheme for Horizontally Fractured Medium

3.3.1 Boundary Conditions

Consider a homogeneous host medium that is discretized by a uniform rectangular grid and contains a horizontal linear slip interface along the numerical grid boundary at depth $Z = (x, z + \frac{1}{2}\Delta z)$. The slip interface vertically separates the homogenous host medium into two spaces such that the upper medium 1 belongs to the region shallower than $Z = (x, z + \frac{1}{2}\Delta z)$ with density ρ_1, S-wave velocity β_1 and P-wave velocity α_1, and the lower medium 2 occupies the region deeper than $Z = (x, z + \frac{1}{2}\Delta z)$ with density ρ_2, S-wave velocity β_2 and P-wave velocity α_2. As discussed in Chap. 2, with the long wavelength assumption and the linear slip theory, this transversely isotropic medium with a vertical asymmetric axis is equivalent to an effective horizontally fractured medium (VTI) (see Table 2.1) in which the fracture is the horizontal infinite linear slip interface at $Z = (x, z + \frac{1}{2}\Delta z)$ and satisfies the nonwelded boundary conditions.

If a horizontal fracture separates a homogeneous isotropic host medium, then $\rho_1 = \rho_2, \beta_1 = \beta_2, \alpha_1 = \alpha_2$ (Fig. 3.2a). This treats the horizontal fracture as the boundary between the evaluated grid points and the nearest-neighbor grid points in the FD stencil, which is mathematically constructed rather than having rock boundaries with impedance contrasts. This model agrees with the real cases that the fractures are often revealed in a uniform lithology, such as coal, shale in the reservoir.

In the FD stencil of the effective horizontally fractured medium shown in Fig. 3.2b, an evaluated displacement rectangle point at (x, z) is surrounded by four grid boundaries but only one z-normal boundary models the fracture and satisfies the imperfectly welded (nonwelded) contact boundary conditions at $Z = \left(x, z + \frac{1}{2}\Delta z\right)$, whereas the rest of the boundaries, two x-normal grid boundaries at $X = (x \pm \frac{1}{2}\Delta x, z)$ and one z-normal boundary at $Z = (x, z - \frac{1}{2}\Delta z)$, satisfy the perfectly welded contact boundary conditions.

The perfectly welded contact boundary conditions at $Z = (x, z - \frac{1}{2}\Delta z)$ are that both displacement and stress are continuous across the welded boundary. Mathematically, that is

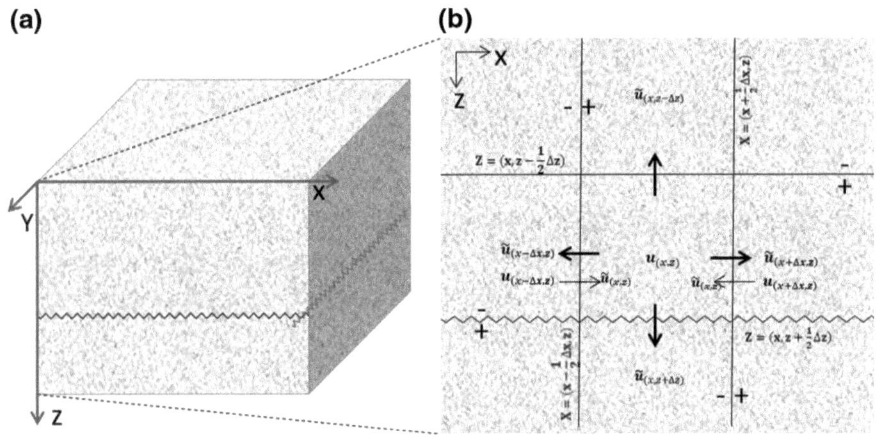

Fig. 3.2 A horizontally fractured medium (**a**) and a finite-difference stencil (**b**) with horizontal fracture in *x*, *z*-domain

$$\begin{cases} u^-_{x(x,z-\frac{1}{2}\Delta z)} = u^+_{x(x,z-\frac{1}{2}\Delta z)} \\ \sigma^-_{xz(x,z-\frac{1}{2}\Delta z)} = \sigma^+_{xz(x,z-\frac{1}{2}\Delta z)} \end{cases}, \tag{3.04a}$$

$$\begin{cases} u^-_{z(x,z-\frac{1}{2}\Delta z)} = u^+_{z(x,z-\frac{1}{2}\Delta z)} \\ \sigma^-_{zz(x,z-\frac{1}{2}\Delta z)} = \sigma^+_{zz(x,z-\frac{1}{2}\Delta z)} \end{cases}, \tag{3.04b}$$

where the subscripts in bracket denote boundary locations, italic letters, e.g. x and z, or xz and zz represent tangential and normal components. Superscripts indicate the sides of the boundary with positive and negative signs.

The imperfectly welded (nonwelded) contact boundary conditions at $Z = \left(x, z + \frac{1}{2}\Delta z\right)$ are where the stress is continuous across the nonwelded boundary, while displacements are discontinuous: the tangential displacement difference is linearly proportional to the tangential stress with the proportionality factor being the tangential fracture compliance S_{T}, while the normal displacement difference is linearly proportional to the normal stress with the proportionality factor being the normal fracture compliance S_{N}. These conditions can be expressed mathematically as

$$\begin{cases} u^+_{x(x,z+\frac{1}{2}\Delta z)} - u^-_{x\left(x,z+\frac{1}{2}\Delta z\right)} = S_{\mathrm{T}}\sigma^+_{xz\left(x,z+\frac{1}{2}\Delta z\right)} \\ \sigma^+_{xz(x,z+\frac{1}{2}\Delta z)} = \sigma^-_{xz(x,z+\frac{1}{2}\Delta z)} \end{cases}, \tag{3.05a}$$

$$\begin{cases} u^+_{z(x,z+\frac{1}{2}\Delta z)} - u^-_{z\left(x,z+\frac{1}{2}\Delta z\right)} = S_{\mathrm{N}}\sigma^+_{zz\left(x,z+\frac{1}{2}\Delta z\right)} \\ \sigma^+_{zz(x,z+\frac{1}{2}\Delta z)} = \sigma^-_{zz(x,z+\frac{1}{2}\Delta z)} \end{cases}, \tag{3.05b}$$

where

$$\sigma^{\pm}_{xz(x,z\pm\frac{1}{2}\Delta z)} = \mu^{\pm}_{(x,z\pm\frac{1}{2}\Delta z)} \left(\frac{\partial u^{\pm}_{x(x,z\pm\frac{1}{2}\Delta z)}}{\partial z} + \frac{\partial u^{\pm}_{z(x,z\pm\frac{1}{2}\Delta z)}}{\partial x} \right), \tag{3.06a}$$

$$\sigma^{\pm}_{zz(x,z\pm\frac{1}{2}\Delta z)} = (\lambda + 2\mu)^{\pm}_{(x,z\pm\frac{1}{2}\Delta z)} \frac{\partial u^{\pm}_{z(x,z\pm\frac{1}{2}\Delta z)}}{\partial z} + \lambda^{\pm}_{(x,z\pm\frac{1}{2}\Delta z)} \frac{\partial u^{\pm}_{x(x,z\pm\frac{1}{2}\Delta z)}}{\partial x}. \tag{3.06b}$$

Since the host medium is a homogeneous medium, the medium properties are invariant with location. That is,

$$\mu^{+}_{(x,z\pm\frac{1}{2}\Delta z)} = \mu_{(x,z\pm\Delta z)} = \mu_{(x,z)} = \mu^{-}_{(x,z\pm\frac{1}{2}\Delta z)}, \tag{3.07}$$

$$\lambda^{+}_{(x,z\pm\frac{1}{2}\Delta z)} = \lambda_{(x,z\pm\Delta z)} = \lambda_{(x,z)} = \lambda^{-}_{(x,z\pm\frac{1}{2}\Delta z)}.. \tag{3.08}$$

The x-normal boundaries at $X = (x \pm \frac{1}{2}\Delta x, z)$ have perfectly welded boundary conditions: displacement and stress are continuous across the boundaries. That is,

$$\begin{cases} u^{+}_{z(x\pm\frac{1}{2}\Delta x,z)} = u^{-}_{z(x\pm\frac{1}{2}\Delta x,z)} \\ \sigma^{+}_{zx(x\pm\frac{1}{2}\Delta x,z)} = \sigma^{-}_{zx(x\pm\frac{1}{2}\Delta x,z)} \end{cases}, \tag{3.09a}$$

$$\begin{cases} u^{+}_{x(x\pm\frac{1}{2}\Delta x,z)} = u^{-}_{x(x\pm\frac{1}{2}\Delta x,z)} \\ \sigma^{+}_{xx(x\pm\frac{1}{2}\Delta x,z)} = \sigma^{-}_{xx(x\pm\frac{1}{2}\Delta x,z)} \end{cases}, \tag{3.09b}$$

where

$$\sigma^{\pm}_{zx(x\pm\frac{1}{2}\Delta x,z)} = \mu^{\pm}_{(x\pm\frac{1}{2}\Delta x,z)} \left(\frac{\partial u^{\pm}_{z(x\pm\frac{1}{2}\Delta x,z)}}{\partial x} + \frac{\partial u^{\pm}_{x(x\pm\frac{1}{2}\Delta x,z)}}{\partial z} \right), \tag{3.10a}$$

$$\sigma^{\pm}_{xx(x\pm\frac{1}{2}\Delta x,z)} = (\lambda + 2\mu)^{\pm}_{(x\pm\frac{1}{2}\Delta x,z)} \frac{\partial u^{\pm}_{x(x\pm\frac{1}{2}\Delta x,z)}}{\partial x} + \lambda^{\pm}_{(x\pm\frac{1}{2}\Delta x,z)} \frac{\partial u^{\pm}_{z(x\pm\frac{1}{2}\Delta x,z)}}{\partial z}. \tag{3.10b}$$

3.3.2 Fictitious Displacement Formulas

In the FD stencil shown in Fig. 3.2b, as explained above, an evaluated point at (x, z) relates to two kinds of boundary conditions that constrain all wave propagation in the horizontally fractured medium. Using the numerical average operator and finite-difference central operator (Appendix B) to expand the fracture boundary conditions in Eqs. (3.05a, 3.05b), the boundary conditions can be expressed in terms of the fictitious displacement and real displacement instead of the generalized displacement and stress items.

$$\begin{cases} \frac{1}{2}\left(u_{x(x,z+\Delta z)}+\tilde{u}_{x(x,z)}\right)-\frac{1}{2}\left(\tilde{u}_{x(x,z+\Delta z)}+u_{x(x,z)}\right) \\ =S_{\mathrm{T}}\mu^{+}_{(x,z+\frac{1}{2}\Delta z)}\left(\frac{u_{z(x+\Delta x,z+\Delta z)}-u_{z(x-\Delta x,z+\Delta z)}}{2\Delta x}+\frac{u_{x(x,z+\Delta z)}-\tilde{u}_{x(x,z)}}{\Delta z}\right), \\ \mu^{+}_{(x,z+\frac{1}{2}\Delta z)}\left(\frac{u_{z(x+\Delta x,z+\Delta z)}-u_{z(x-\Delta x,z+\Delta z)}}{2\Delta x}+\frac{u_{x(x,z+\Delta z)}-\tilde{u}_{x(x,z)}}{\Delta z}\right) \\ =\mu^{-}_{(x,z+\frac{1}{2}\Delta z)}\left(\frac{u_{z(x+\Delta x,z)}-u_{z(x-\Delta x,z)}}{2\Delta x}+\frac{\tilde{u}_{x(x,z+\Delta z)}-u_{x(x,z)}}{\Delta z}\right), \end{cases} \tag{3.11a}$$

$$\begin{cases} \frac{1}{2}\left(u_{z(x,z+\Delta z)}+\tilde{u}_{z(x,z)}\right)-\frac{1}{2}\left(\tilde{u}_{z(x,z+\Delta z)}+u_{z(x,z)}\right) \\ =\lambda^{+}_{(x,z+\frac{1}{2}\Delta z)}\left(\frac{u_{x(x+\Delta x,z+\Delta z)}-u_{x(x-\Delta x,z+\Delta z)}}{2\Delta x}\right)+(\lambda+2\mu)^{+}_{(x,z+\frac{1}{2}\Delta z)}\left(\frac{u_{z(x,z+\Delta z)}-\tilde{u}_{z(x,z)}}{\Delta z}\right), \\ \lambda^{+}_{(x,z+\frac{1}{2}\Delta z)}\left(\frac{u_{x(x+\Delta x,z+\Delta z)}-u_{x(x-\Delta x,z+\Delta z)}}{2\Delta x}\right)+(\lambda+2\mu)^{+}_{(x,z+\frac{1}{2}\Delta z)}\left(\frac{u_{z(x,z+\Delta z)}-\tilde{u}_{z(x,z)}}{\Delta z}\right) \\ =\lambda^{-}_{(x,z+\frac{1}{2}\Delta z)}\left(\frac{u_{x(x+\Delta x,z)}-u_{x(x-\Delta x,z)}}{2\Delta x}\right)+(\lambda+2\mu)^{-}_{(x,z+\frac{1}{2}\Delta z)}\left(\frac{\tilde{u}_{z(x,z+\Delta z)}-u_{z(x,z)}}{\Delta z}\right). \end{cases} \tag{3.11b}$$

Equations (3.11a), (3.11b) are the fracture boundary conditions in terms of real and fictitious displacements for the tangential component (x) and the normal component (z), respectively. The fictitious displacements in the equation tend to extend the media into each other in order to present the displacement difference of the two sides of the fracture. To derive Eqs. (3.11a, 3.11b), the normal derivatives of the displacements in the x, z-components are approximated to $O\,(\Delta z)^2$ (Appendix B), the tangential derivative of the displacements in x, z-components are approximated to $O\left((\Delta x)^2, \Delta z\right)$. For example,

$$\frac{\partial \boldsymbol{u}^{+}_{(x,z+\frac{1}{2}\Delta z)}}{\partial z}=\frac{\boldsymbol{u}_{(x,z+\Delta z)}-\tilde{\boldsymbol{u}}_{(x,z)}}{\Delta z}, \tag{3.12a}$$

$$\frac{\partial \boldsymbol{u}^{+}_{(x,z+\frac{1}{2}\Delta z)}}{\partial x}=\frac{\boldsymbol{u}_{(x+\Delta x,z+\Delta z)}-\boldsymbol{u}_{(x-\Delta x,z+\Delta z)}}{2\Delta x}. \tag{3.12b}$$

With those approximations, the next-nearest-neighbor fictitious points need to be eliminated from the boundary conditions expressions to avoid the underdetermined problems (Slawinski 1999). Equations (3.11a, 3.11b) present a linear system that can be simultaneously solved for the four expressions for the four unknown fictitious displacement formulae:

$$\begin{aligned} \tilde{u}_{x(x,z)}=&\frac{1}{1+\varepsilon_{\mathrm{T}}}\left[u_{x(x,z)}+\varepsilon_{\mathrm{T}}u_{x(x,z+\Delta z)}-\frac{1}{4}\left(u_{z(x+\Delta x,z)}-u_{z(x-\Delta x,z)}\right)\right.\\ &\left.+\frac{1}{4}(1+2\varepsilon_{\mathrm{T}})\left(u_{z(x+\Delta x,z+\Delta z)}-u_{z(x-\Delta x,z+\Delta z)}\right)\right], \end{aligned} \tag{3.13a}$$

$$\tilde{u}_{x(x,z+\Delta z)} = \frac{1}{1+\varepsilon_{\mathrm{T}}}\Big[u_{x(x,z+\Delta z)} + \varepsilon_{\mathrm{T}} u_{x(x,z)} - \frac{1}{4}(1+2\varepsilon_{\mathrm{T}})\big(u_{z(x+\Delta x,z)} - u_{z(x-\Delta x,z)}\big) + \frac{1}{4}\big(u_{z(x+\Delta x,z+\Delta z)} - u_{z(x-\Delta x,z+\Delta z)}\big)\Big], \tag{3.13b}$$

$$\tilde{u}_{z(x,z)} = \frac{1}{1+\delta_{\mathrm{N}}}\Big[u_{z(x,z)} + \delta_{\mathrm{N}} u_{z(x,z+\Delta z)} - \frac{1}{4}\frac{\gamma_{\mathrm{N}}}{\delta_{\mathrm{N}}}\big(u_{x(x+\Delta x,z)} - u_{x(x-\Delta x,z)}\big) + \frac{1}{4}(1+2\delta_{\mathrm{N}})\frac{\gamma_{\mathrm{N}}}{\delta_{\mathrm{N}}}\big(u_{x(x+\Delta x,z+\Delta z)} - u_{x(x-\Delta x,z+\Delta z)}\big)\Big], \tag{3.13c}$$

$$\tilde{u}_{z(x,z+\Delta z)} = \frac{1}{1+\delta_{\mathrm{N}}}\Big[u_{z(x,z+\Delta z)} + \delta_{\mathrm{N}} u_{z(x,z)} - \frac{1}{4}(1+2\delta_{\mathrm{N}})\frac{\gamma_{\mathrm{N}}}{\delta_{\mathrm{N}}}\big(u_{x(x+\Delta x,z)} - u_{x(x-\Delta x,z)}\big) + \frac{1}{4}\frac{\gamma_{\mathrm{N}}}{\delta_{\mathrm{N}}}\big(u_{x(x+\Delta x,z+\Delta z)} - u_{x(x-\Delta x,z+\Delta z)}\big)\Big]. \tag{3.13d}$$

Equations (3.13a–3.13d) are formulae for the tangential and normal fictitious displacements points at (x, z) and $(x, z+\Delta z)$ that relate to the z-normal fracture interface. The equations show two representations (real and fictitious) of the displacement at the same grid point presenting the displacement difference between the real and fictitious point that not only relates material discontinuities in the nearest grid point, but also involves a material discontinuity in the next-nearest-neighbor points. In Eqs. (3.13a–3.13d),

$$\varepsilon_{\mathrm{T}} = \frac{S_{\mathrm{T}}\mu}{\Delta d}, \quad \delta_{\mathrm{N}} = \frac{S_{\mathrm{N}}(\lambda+2\mu)}{\Delta d}, \quad \gamma_{\mathrm{N}} = \frac{S_{\mathrm{N}}\lambda}{\Delta d}, \tag{3.14}$$

where $\Delta d = \Delta x = \Delta z$. $\varepsilon_{\mathrm{T}}, \delta_{\mathrm{N}}$ and γ_{N} are the dimensionless nonweldedness parameters that depend on the medium properties, fracture compliances and grid parameters. ε_{T} determines the contribution to the tangential displacement (u_x) discontinuity across the fracture interface from the continuous change in the displacement. Similarly, δ_{N} and γ_{N} determine the contribution to the normal displacement (u_z) discontinuity across the fracture interface from the continuous change in the displacement (Slawinski 1999, p. 208). The fictitious displacement formulae in Eqs. (3.13a–3.13d) can be rewritten in a compact way as

$$\tilde{\boldsymbol{u}}_{(x,z)_{\text{nonwelded}}} = N\boldsymbol{u}_{(x,z)} + M\boldsymbol{u}_{(x,z+\Delta d)} - \frac{1}{4}E\big(\boldsymbol{u}_{(x-\Delta d,z)} - \boldsymbol{u}_{(x+\Delta d,z)}\big) + \frac{1}{4}G\big(\boldsymbol{u}_{(x-\Delta d,z+\Delta d)} - \boldsymbol{u}_{(x+\Delta d,z+\Delta d)}\big), \tag{3.15}$$

$$\tilde{\boldsymbol{u}}_{(x,z+\Delta d)_{\text{nonwelded}}} = N\boldsymbol{u}_{(x,z+\Delta d)} + M\boldsymbol{u}_{(x,z)} - \frac{1}{4}G\left(\boldsymbol{u}_{(x+\Delta d,z)} - \boldsymbol{u}_{(x-\Delta d,z)}\right) + \frac{1}{4}E\left(\boldsymbol{u}_{(x+\Delta d,z+\Delta d)} - \boldsymbol{u}_{(x-\Delta d,z+\Delta d)}\right), \tag{3.16}$$

where

$$N = \begin{bmatrix} \frac{1}{1+\varepsilon_T} & 0 \\ 0 & \frac{1}{1+\delta_N} \end{bmatrix}, \tag{3.17a}$$

$$M = \begin{bmatrix} \frac{\varepsilon_T}{1+\varepsilon_T} & 0 \\ 0 & \frac{\delta_N}{1+\delta_N} \end{bmatrix}, \tag{3.17b}$$

$$G = \begin{bmatrix} 0 & \frac{1+2\varepsilon_T}{1+\varepsilon_T} \\ \frac{\gamma_N(1+2\delta_N)}{\delta_N(1+\delta_N)} & 0 \end{bmatrix}, \tag{3.17c}$$

$$E = \begin{bmatrix} 0 & \frac{1}{1+\varepsilon_T} \\ \frac{\gamma_N}{\delta_N(1+\delta_N)} & 0 \end{bmatrix}. \tag{3.17d}$$

If the dimensionless nonweldedness parameters vanish as $S_T = 0, S_N = 0$. The fracture boundary changes to a z-normal perfectly welded boundary at $Z = \left(x, z + \frac{1}{2}\Delta z\right)$. So Eqs. (3.15, 3.16) can be modified as

$$\tilde{\boldsymbol{u}}_{(x,z)} = \boldsymbol{u}_{(x,z)} + \frac{1}{4}H\left(\boldsymbol{u}_{(x+\Delta d,z+\Delta d)} - \boldsymbol{u}_{(x-\Delta d,z+\Delta d)} - \boldsymbol{u}_{(x+\Delta d,z)} + \boldsymbol{u}_{(x-\Delta d,z)}\right), \tag{3.18}$$

$$\tilde{\boldsymbol{u}}_{(x,z+\Delta d)} = \boldsymbol{u}_{(x,z+\Delta d)} + \frac{1}{4}H\left(\boldsymbol{u}_{(x+\Delta d,z+\Delta d)} - \boldsymbol{u}_{(x-\Delta d,z+\Delta d)} - \boldsymbol{u}_{(x+\Delta d,z)} + \boldsymbol{u}_{(x-\Delta d,z)}\right) \tag{3.19}$$

Slawinski (1999) presented a simple method to determine the fictitious displacement for the nearest-neighbor grid points from the central fictitious displacement grid point. For example, the fictitious displacement at $\tilde{\boldsymbol{u}}_{(x,z-\Delta d)}$ can be computed from $\tilde{\boldsymbol{u}}_{(x,z)}$ in Eq. (3.18) by shifting "z" to "$z - \Delta d$". Thus,

$$\tilde{\boldsymbol{u}}_{(x,z-\Delta d)} = \boldsymbol{u}_{(x,z-\Delta d)} + \frac{1}{4}H\left(\boldsymbol{u}_{(x+\Delta d,z)} - \boldsymbol{u}_{(x-\Delta d,z)} - \boldsymbol{u}_{(x+\Delta d,z-\Delta d)} + \boldsymbol{u}_{(x-\Delta d,z-\Delta d)}\right) \tag{3.20}$$

Computing the fictitious displacements at $\tilde{\boldsymbol{u}}_{(x+\Delta d,z)}$ and $\tilde{\boldsymbol{u}}_{(x-\Delta d,z)}$ can be simply obtained from $\tilde{\boldsymbol{u}}_{(x,z+\Delta d)}$ and $\tilde{\boldsymbol{u}}_{(x,z-\Delta d)}$ by exchanging "x" to "z" (Slawinski 1999). Then we have

$$\tilde{\boldsymbol{u}}_{(x-\Delta d,z)} = \boldsymbol{u}_{(x-\Delta d,z)} + \frac{1}{4}H^{\mathrm{T}}\left(\boldsymbol{u}_{(x,z+\Delta d)} - \boldsymbol{u}_{(x,z-\Delta d)} - \boldsymbol{u}_{(x-\Delta d,z+\Delta d)} + \boldsymbol{u}_{(x-\Delta d,z-\Delta d)}\right), \tag{3.21}$$

$$\tilde{\boldsymbol{u}}_{(x+\Delta d,z)} = \boldsymbol{u}_{(x+\Delta d,z)} + \frac{1}{4}H^{\mathrm{T}}\left(\boldsymbol{u}_{(x+\Delta d,z+\Delta d)} - \boldsymbol{u}_{(x+\Delta d,z-\Delta d)} - \boldsymbol{u}_{(x,z+\Delta d)} + \boldsymbol{u}_{(x,z-\Delta d)}\right), \tag{3.22}$$

where

$$H = \begin{bmatrix} 0 & 1 \\ \frac{\lambda}{\lambda+2\mu} & 0 \end{bmatrix}, \quad H^{\mathrm{T}} = \begin{bmatrix} 0 & \frac{\lambda}{\lambda+2\mu} \\ 1 & 0 \end{bmatrix}. \tag{3.23}$$

Equations (3.19)–(3.22) are fictitious displacement formulae for the nearest-neighbor grid points related to central grid point with the welded boundary conditions. The fictitious displacement formulae express the difference between fictitious displacement and real displacement that is caused by the continuity of stress boundary conditions, which couple two components (tangential and normal components) displacements along the boundary (Slawinski 1999).

3.3.3 Equation of Motion

A P-SV wave equation of motion in a homogeneous isotropic medium may be written as (Aki and Richards 1980)

$$\frac{\partial^2 \boldsymbol{u}}{\partial t^2} = A\frac{\partial^2 \boldsymbol{u}}{\partial x^2} + B\frac{\partial^2 \boldsymbol{u}}{\partial x \partial z} + C\frac{\partial^2 \boldsymbol{u}}{\partial z^2}, \tag{3.24}$$

and

$$A = \begin{pmatrix} \alpha^2 & 0 \\ 0 & \beta^2 \end{pmatrix}, \quad B = \begin{pmatrix} 0 & \alpha^2 - \beta^2 \\ \alpha^2 - \beta^2 & 0 \end{pmatrix}, \quad C = \begin{pmatrix} \beta^2 & 0 \\ 0 & \alpha^2 \end{pmatrix}, \tag{3.25}$$

where α and β are, respectively, the compressional and shear wave velocities. Using the second-order central finite-difference approximation to solve the wave Eq. (3.24) at point $\boldsymbol{u}_{(x,z)}$ (Aki and Richards 1980), one obtains:

$$
\begin{aligned}
\boldsymbol{u}_{(x,z)}^{t+1} = & -\boldsymbol{u}_{(x,z)}^{t-1} + 2\boldsymbol{u}_{(x,z)}^{t} + \left(\frac{\Delta t}{\Delta d}\right)^2 \mathrm{A}\left(\boldsymbol{u}_{(x+\Delta d,z)}^{t} - 2\boldsymbol{u}_{(x,z)}^{t} + \boldsymbol{u}_{(x-\Delta d,z)}^{t}\right) \\
& + \frac{(\Delta t)^2}{4(\Delta d)^2}\mathrm{B}\left(\boldsymbol{u}_{(x+\Delta d,z+\Delta d)}^{t} - \boldsymbol{u}_{(x+\Delta d,z-\Delta d)}^{t} - \boldsymbol{u}_{(x-\Delta d,z+\Delta d)}^{t} + \boldsymbol{u}_{(x-\Delta d,z-\Delta d)}^{t}\right) \\
& + \left(\frac{\Delta t}{\Delta d}\right)^2 \mathrm{C}\left(\boldsymbol{u}_{(z,z+\Delta d)}^{t} - 2\boldsymbol{u}_{(x,z)}^{t} + \boldsymbol{u}_{(x,z-\Delta d)}^{t}\right).
\end{aligned}
\tag{3.26}
$$

The subscripts in parentheses denote the grid locations. Clearly, for the P-SV FD stencil in Eq. (3.26), an evaluation of $\boldsymbol{u}_{(x,z)}$ involves four nearest-neighbor grid point $\boldsymbol{u}_{(x\pm d,z)}$ and $\boldsymbol{u}_{(x,z\pm d)}$, and four next-nearest-neighbor grid points $\boldsymbol{u}_{(x\pm\Delta x,z\pm\Delta z)}$ (Slawinski and Krebes 2002). The polynomial of the next-nearest-neighbor points, however, can be reduced by taking Taylor expansion in the first order approximation.

$$
\begin{aligned}
& \boldsymbol{u}_{(x+\Delta d,z+\Delta d)}^{t} - \boldsymbol{u}_{(x+\Delta d,z-\Delta d)}^{t} - \boldsymbol{u}_{(x-\Delta d,z+\Delta d)}^{t} + \boldsymbol{u}_{(x-\Delta d,z-\Delta d)}^{t} \\
& \approx \left(\boldsymbol{u}_{(x,z)} + \frac{\partial \boldsymbol{u}_{(x,z)}}{\partial x}\Delta d + \frac{\partial \boldsymbol{u}_{(x,z)}}{\partial z}\Delta d\right) - \left(\boldsymbol{u}_{(x,z)} + \frac{\partial \boldsymbol{u}_{(x,z)}}{\partial x}\Delta d - \frac{\partial \boldsymbol{u}_{(x,z)}}{\partial z}\Delta d\right) \\
& \quad - \left(\boldsymbol{u}_{(x,z)} - \frac{\partial \boldsymbol{u}_{(x,z)}}{\partial x}\Delta d + \frac{\partial \boldsymbol{u}_{(x,z)}}{\partial z}\Delta d\right) + \left(\boldsymbol{u}_{(x,z)} - \frac{\partial \boldsymbol{u}_{(x,z)}}{\partial x}\Delta d - \frac{\partial \boldsymbol{u}_{(x,z)}}{\partial z}\Delta d\right) \\
& \approx 0.
\end{aligned}
\tag{3.27}
$$

Consequently, the equation of motion (3.26) can be approximated and the evaluated point $\boldsymbol{u}_{(x,z)}$ only relates to the four nearest-neighbor grid points.

$$
\begin{aligned}
\boldsymbol{u}_{(x,z)}^{t+1} = & -\boldsymbol{u}_{(x,z)}^{t-1} + 2\boldsymbol{u}_{(x,z)}^{t} \\
& + \left(\frac{\Delta t}{\Delta d}\right)^2 \mathrm{A}\left(\boldsymbol{u}_{(x+\Delta d,z)}^{t} - 2\boldsymbol{u}_{(x,z)}^{t} + \boldsymbol{u}_{(x-\Delta d,z)}^{t}\right) \\
& + \left(\frac{\Delta t}{\Delta d}\right)^2 \mathrm{C}\left(\boldsymbol{u}_{(x,z+\Delta d)}^{t} - 2\boldsymbol{u}_{(x,z)}^{t} + \boldsymbol{u}_{(x,z-\Delta d)}^{t}\right).
\end{aligned}
\tag{3.28}
$$

In order to explicitly impose the boundary conditions at $Z = (x, z \pm \frac{1}{2}\Delta d)$ and $X = (x \pm \frac{1}{2}\Delta d, z)$, the real displacement at the nearest-neighbor grid in Eq. (3.28) should be replaced by corresponding fictitious displacement. The resulting equation should be

$$\begin{aligned}\boldsymbol{u}^{t+1}_{(x,z)} &= -\boldsymbol{u}^{t-1}_{(x,z)} + 2\boldsymbol{u}^{t}_{(x,z)} \\ &+ \left(\frac{\Delta t}{\Delta d}\right)^2 \mathrm{A}\left(\tilde{\boldsymbol{u}}^{t}_{(x+\Delta d,z)} - 2\boldsymbol{u}^{t}_{(x,z)} + \tilde{\boldsymbol{u}}^{t}_{(x-\Delta d,z)}\right) \\ &+ \left(\frac{\Delta t}{\Delta d}\right)^2 \mathrm{C}\left(\tilde{\boldsymbol{u}}^{t}_{(x,z+\Delta d)_{\text{nonwelded}}} - 2\boldsymbol{u}^{t}_{(x,z)} + \tilde{\boldsymbol{u}}^{t}_{(x,z-\Delta d)}\right).\end{aligned} \tag{3.29}$$

In Eq. (3.29), the fictitious displacement with subscript "nonwelded", $\tilde{\boldsymbol{u}}^{t}_{(x,z+\Delta d)_{\text{nonwelded}}}$, is determined by the nonwelded boundary conditions in Eq. (3.16), and is distinct from the other fictitious displacement $\tilde{\boldsymbol{u}}_{(x\pm\Delta d,z)}, \tilde{\boldsymbol{u}}_{(x,z-\Delta d)}$ determined by the welded boundary conditions. Substituting those four fictitious displacements into Eq. (3.29) yields a P-SV equation of motion for a horizontally fractured homogeneous medium:

$$\begin{aligned}\boldsymbol{u}^{t+1}_{(x,z)} &= -\boldsymbol{u}^{t-1}_{(x,z)} + 2\boldsymbol{u}^{t}_{(x,z)} \\ &+ \left(\frac{\Delta t}{\Delta d}\right)^2 A\left[\left(\boldsymbol{u}^{t}_{(x+\Delta d,z)} - 2\boldsymbol{u}^{t}_{(x,z)} + \boldsymbol{u}^{t}_{(x-\Delta d,z)}\right)\right] \\ &+ \frac{1}{4}\left(\frac{\Delta t}{\Delta d}\right)^2 B\left[\boldsymbol{u}^{t}_{(x+\Delta d,z+\Delta d)} - \boldsymbol{u}^{t}_{(x-\Delta d,z+\Delta d)} - \boldsymbol{u}^{t}_{(x+\Delta d,z-\Delta d)} + \boldsymbol{u}^{t}_{(x-\Delta d,z-\Delta d)}\right] \\ &+ \left(\frac{\Delta t}{\Delta d}\right)^2 C\left[\boldsymbol{u}^{t}_{(x,z+\Delta d)} - 2\boldsymbol{u}^{t}_{(x,z)} + \boldsymbol{u}^{t}_{(x,z-\Delta d)}\right] \\ &+ \left(\frac{\Delta t}{\Delta d}\right)^2 \mathrm{CM}\left[\boldsymbol{u}^{t}_{(x,z)} - \boldsymbol{u}^{t}_{(x,z+\Delta d)} - \frac{1}{4}H\left(\boldsymbol{u}^{t}_{(x+\Delta d,z+\Delta d)} - \boldsymbol{u}^{t}_{(x-\Delta d,z+\Delta d)}\right.\right. \\ &\left.\left.+ \boldsymbol{u}^{t}_{(x+\Delta d,z)} - \boldsymbol{u}^{t}_{(x-\Delta d,z)}\right)\right].\end{aligned} \tag{3.30}$$

Note that the four next-nearest-neighbor displacements (mixed derivative items) have been taken from the fictitious displacement formulae. Coefficient M is only associated with the particular fractured boundary and are dependent on whether material is discontinuity. Equation (3.30) can be used to model a set of horizontal fractures in the fractured medium with impedance contrast.

3.4 Finite-Difference Scheme for Vertically Fractured Medium

3.4.1 Boundary Conditions

Consider a medium that has a linear slip interface vertically inserted into a homogeneous host medium at $X = \left(x + \frac{1}{2}\Delta x, z\right)$ (Fig. 3.3). The slip interface

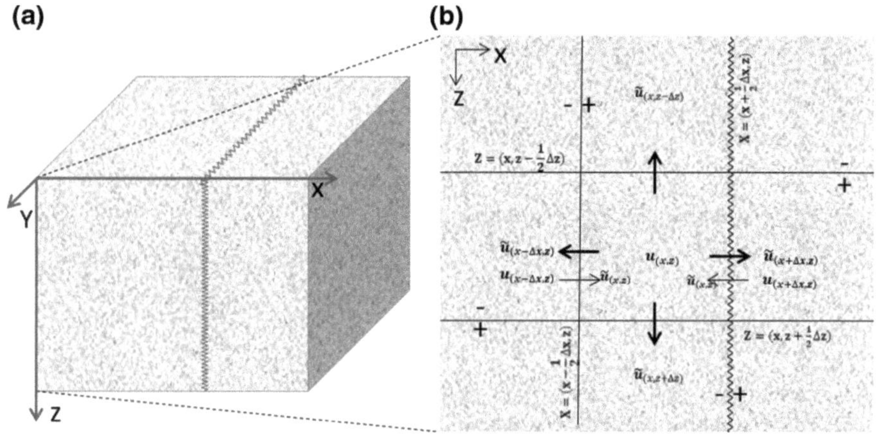

Fig. 3.3 Vertically fractured medium (**a**) and a FD stencil (**b**) with vertical fracture in x, z-domain

horizontally separates the homogenous host medium into two spaces, of which medium 1 belongs to the region that is horizontally smaller than $X = \left(x + \frac{1}{2}\Delta x, z\right)$ with density ρ_1, S-wave velocity β_1 and P-wave velocity α_1 and medium 2 occupies the region in the right side of the interface $X = \left(x + \frac{1}{2}\Delta x, z\right)$ with density ρ_2, S-wave velocity β_2 and P-wave velocity α_2. If $\rho_1 = \rho_2, \beta_1 = \beta_2, \alpha_1 = \alpha_2$, the host medium is a homogeneous isotropic medium. Following the long wavelength assumption and the linear slip theory, this reconstructed transversely isotropic medium with a horizontal symmetry axis is equivalent to an effective vertically fractured medium. One symmetric plane coincides with the vertical fracture plane and is parallel to $[y, z]-$ plane in the coordinate system (perpendicular to the x-axis) (Fig. 3.3a). Similarly to the effective horizontally fractured medium, an evaluated central rectangle point at (x, z) should encounter four grid boundaries at $X = \left(x \pm \frac{1}{2}\Delta x, z\right)$ and $Z = \left(x, z \pm \frac{1}{2}\Delta z\right)$. However, the boundary characteristics at $X = \left(x + \frac{1}{2}\Delta x, z\right)$ and $Z = \left(x, z + \frac{1}{2}\Delta z\right)$ are contrary to the horizontally fractured medium case: the boundary at $X = \left(x + \frac{1}{2}\Delta x, z\right)$ is a fractured nonwelded contact boundary, whereas the boundary at $Z = \left(x, z + \frac{1}{2}\Delta z\right)$ changes to a perfect welded contact boundary. This means that referring to the evaluated point (x, z), the boundaries at $Z = \left(x, z \pm \frac{1}{2}\Delta z\right)$ and $X = \left(x - \frac{1}{2}\Delta x, z\right)$ satisfy the perfectly welded contact boundary conditions, but the boundary at $X = \left(x + \frac{1}{2}\Delta x, z\right)$ will have imperfectly welded (nonwelded) contact boundary conditions. The z-normal boundaries can be expressed as

$$\begin{cases} u^{+}_{x(x,z\pm\frac{1}{2}\Delta z)} = u^{-}_{x(x,z\pm\frac{1}{2}\Delta z)} \\ \sigma^{+}_{xz(x,z\pm\frac{1}{2}\Delta z)} = \sigma^{-}_{xz(x,z\pm\frac{1}{2}\Delta z)} \end{cases}, \tag{3.31a}$$

$$\begin{cases} u^{+}_{z(x,z\pm\frac{1}{2}\Delta z)} = u^{-}_{z(x,z\pm\frac{1}{2}\Delta z)} \\ \sigma^{+}_{zz(x,z\pm\frac{1}{2}\Delta z)} = \sigma^{-}_{zz(x,z\pm\frac{1}{2}\Delta z)} \end{cases}, \tag{3.31b}$$

where

$$\sigma^{\pm}_{xz(x,z\pm\frac{1}{2}\Delta z)} = \mu^{\pm}_{(x,z\pm\frac{1}{2}\Delta z)} \left(\frac{\partial u^{\pm}_{x(x,z\pm\frac{1}{2}\Delta z)}}{\partial z} + \frac{\partial u^{\pm}_{z(x,z\pm\frac{1}{2}\Delta z)}}{\partial x} \right), \tag{3.32a}$$

$$\sigma^{\pm}_{zz(x,z\pm\frac{1}{2}\Delta z)} = (\lambda + 2\mu)^{\pm}_{(x,z\pm\frac{1}{2}\Delta z)} \frac{\partial u^{\pm}_{z(x,z\pm\frac{1}{2}\Delta z)}}{\partial z} + \lambda^{\pm}_{(x,z\pm\frac{1}{2}\Delta z)} \frac{\partial u^{\pm}_{x(x,z\pm\frac{1}{2}\Delta z)}}{\partial x}. \tag{3.32b}$$

Equations (3.31a, 3.31b) indicate that the z-normal boundaries at $Z = (x, z \pm \frac{1}{2}\Delta z)$ satisfy welded boundary conditions in which both displacement and stress are continuous across the boundaries.

For the x-normal boundaries at $Z = (x + \frac{1}{2}\Delta x, z)$, we have

$$\begin{cases} u^{+}_{x(x+\frac{1}{2}\Delta x,z)} - u^{-}_{x(x+\frac{1}{2}\Delta x,z)} = S_{\mathrm{N}}\sigma^{+}_{xx(x+\frac{1}{2}\Delta x,z)} \\ \sigma^{+}_{xx(x+\frac{1}{2}\Delta x,z)}(xx) = \sigma^{-}_{xx(x+\frac{1}{2}\Delta x,z)}(xx)) \end{cases}, \tag{3.33a}$$

$$\begin{cases} u^{+}_{z(x+\frac{1}{2}\Delta x,z)} - u^{-}_{z(x+\frac{1}{2}\Delta x,z)} = S_{\mathrm{T}}\sigma^{+}_{zx(x+\frac{1}{2}\Delta x,z)} \\ \sigma^{-}_{zx(x+\frac{1}{2}\Delta x,z)} = \sigma^{+}_{zx(x+\frac{1}{2}\Delta x,z)} \end{cases}. \tag{3.33b}$$

Equations (3.33a, 3.33b) indicated that the vertical fracture interface at $Z = (x + \frac{1}{2}\Delta x, z)$ satisfies nonwelded boundary conditions for which the stress is continuous across the fracture, but the displacements are discontinuous across the fracture. And the difference in tangential displacements is linearly proportional to the normal stress with the proportionality constant being the normal fracture compliance S_{N}, while the difference in normal displacements is linearly proportional to the tangential stress with the proportionality constant being the tangential fracture compliance S_{T}. Another x-normal boundary with perfectly welded boundary conations at $Z = (x - \frac{1}{2}\Delta x, z)$ is

$$\begin{cases} u^{+}_{z(x-\frac{1}{2}\Delta x,z)} = u^{-}_{z(x-\frac{1}{2}\Delta x,z)} \\ \sigma^{+}_{zx(x-\frac{1}{2}\Delta x,z)} = \sigma^{-}_{zx(x-\frac{1}{2}\Delta x,z)} \end{cases}, \tag{3.34a}$$

$$\begin{cases} u^{+}_{x(x-\frac{1}{2}\Delta x,z)} = u^{-}_{x(x-\frac{1}{2}\Delta x,z)} \\ \sigma^{+}_{xx(x-\frac{1}{2}\Delta x,z)} = \sigma^{-}_{xx(x-\frac{1}{2}\Delta x,z)} \end{cases}, \tag{3.34b}$$

where

$$\sigma^{\pm}_{zx(x\pm\frac{1}{2}\Delta x,z)} = \mu^{\pm}_{(x\pm\frac{1}{2}\Delta x,z)}\left(\frac{\partial u_z^{\pm}}{\partial x} + \frac{\partial u_x^{\pm}}{\partial z}\right), \tag{3.35a}$$

$$\sigma^{\pm}_{xx(x\pm\frac{1}{2}\Delta x,z)} = (\lambda + 2\mu)^{\pm}_{(x\pm\frac{1}{2}\Delta x,z)}\frac{\partial u_x^{\pm}}{\partial x} + \lambda^{\pm}_{(x\pm\frac{1}{2}\Delta x,z)}\frac{\partial u_z^{\pm}}{\partial z}. \tag{3.35b}$$

Equation (3.34a, 3.34b) states that the displacement and stress are continuous across the boundary at $X = \left(x - \frac{1}{2}\Delta x, z\right)$.

3.4.2 Fictitious Displacement Formulas

Following the horizontal fracture boundary procedure to expand the fracture boundary conditions in Eq. (3.33a, 3.33b) in terms of the fictitious displacement and real displacement at the points, one may eliminate the generalized displacement and stress terms in the boundary condition at the fracture as follows:

$$\begin{cases} \frac{1}{2}\left(u_{x(x+\Delta x,z)} + \tilde{u}_{x(x,z)}\right) - \frac{1}{2}\left(\tilde{u}_{x(x+\Delta x,z)} + u_{x(x,z)}\right) \\ = S_{\mathrm{N}}\left[\lambda^{+}_{(x+\frac{1}{2}\Delta x,z)}\left(\frac{u_{z(x+\Delta x,z+\Delta z)} - u_{z(x+\Delta x,z-\Delta z)}}{2\Delta z}\right) + (\lambda+2\mu)^{+}_{(x+\frac{1}{2}\Delta x,z)}\left(\frac{u_{x(x+\Delta x,z)} - \tilde{u}_{x(x,z)}}{\Delta x}\right)\right], \\ \lambda^{+}_{(x+\frac{1}{2}\Delta x,z)}\left(\frac{u_{z(x+\Delta x,z+\Delta z)} - u_{z(x+\Delta x,z-\Delta z)}}{2\Delta z}\right) + (\lambda+2\mu)^{+}_{(x+\frac{1}{2}\Delta x,z)}\left(\frac{u_{x(x+\Delta x,z)} - \tilde{u}_{x(x,z)}}{\Delta x}\right) \\ = \lambda^{-}_{(x+\frac{1}{2}\Delta x,z)}\left(\frac{u_{z(x,z+\Delta z)} - u_{z(x,z-\Delta z)}}{2\Delta z}\right) + (\lambda+2\mu)^{-}_{(x+\frac{1}{2}\Delta x,z)}\left(\frac{\tilde{u}_{x(x+\Delta x,z)} - u_{x(x,z)}}{\Delta x}\right), \end{cases}, \tag{3.36a}$$

$$\begin{cases} \frac{1}{2}\left(u_{z(x+\Delta x,z)} + \tilde{u}_{z(x,z)}\right) - \frac{1}{2}\left(\tilde{u}_{z(x+\Delta x,z)} + u_{z(x,z)}\right) \\ = S_{\mathrm{T}}\mu^{+}_{(x+\frac{1}{2}\Delta x,z)}\left(\frac{u_{x(x+\Delta x,z+\Delta z)} - u_{x(x+\Delta x,z-\Delta z)}}{2\Delta z} + \frac{u_{z(x+\Delta x,z)} - \tilde{u}_{z(x,z)}}{\Delta x}\right), \\ \mu^{+}_{(x+\frac{1}{2}\Delta x,z)}\left(\frac{u_{x(x+\Delta x,z+\Delta z)} - u_{x(x+\Delta x,z-\Delta z)}}{2\Delta z} + \frac{u_{z(x+\Delta x,z)} - \tilde{u}_{z(x,z)}}{\Delta x}\right) \\ = \mu^{-}_{(x+\frac{1}{2}\Delta x,z)}\left(\frac{u_{x(x,z+\Delta z)} - u_{x(x,z-\Delta z)}}{2\Delta x} + \frac{\tilde{u}_{z(x+\Delta x,z)} - u_{z(x,z)}}{\Delta z}\right). \end{cases}, \tag{3.36b}$$

Using the similar approximation as Eqs. (3.12a, 3.12b) as $O\left(\Delta x, (\Delta z)^2\right)$, one may solve Eqs. (3.36a, 3.36b) for the four unknown fictitious displacements in terms of real displacements as

$$\begin{aligned} \tilde{u}_{x(x,z)} = & \frac{1}{1+\delta_{\mathrm{N}}}\left[u_{x(x,z)} + \delta_{\mathrm{N}} u_{x(x+\Delta x,z)} - \frac{1}{4}\frac{\gamma_{\mathrm{N}}}{\delta_{\mathrm{N}}}\left(u_{z(x,z-\Delta z)} - u_{z(x,z+\Delta z)}\right)\right. \\ & \left.+ \frac{1}{4}(1+2\delta_{\mathrm{N}})\frac{\gamma_{\mathrm{N}}}{\delta_{\mathrm{N}}}\left(u_{z(x+\Delta x,z-\Delta z)} - u_{z(x+\Delta x,z+\Delta z)}\right)\right]. \end{aligned} \tag{3.37a}$$

$$\tilde{u}_{x(x+\Delta x,z)} = \frac{1}{1+\delta_{\mathrm{N}}}\left[u_{x(x+\Delta x,z)} + \delta_{\mathrm{N}} u_{x(x,z)} - \frac{1}{4}(1+2\delta)\frac{\gamma_{\mathrm{N}}}{\delta_{\mathrm{N}}}\left(u_{z(x,z+\Delta z)} - u_{z(x,z-\Delta z)}\right) + \frac{1}{4}\frac{\gamma_{\mathrm{N}}}{\delta_{\mathrm{N}}}\left(u_{z(x+\Delta x,z+\Delta z)} - u_{z(x+\Delta x,z-\Delta z)}\right)\right] \tag{3.37b}$$

$$\tilde{u}_{z(x,z)} = \frac{1}{1+\varepsilon_{\mathrm{T}}}\left[u_{z(x,z)} + \varepsilon_{\mathrm{T}} u_{z(x+\Delta x,z)} - \frac{1}{4}\left(u_{x(x,z-\Delta z)} - u_{x(x,z+\Delta z)}\right) + \frac{1}{4}(1+2\varepsilon_{\mathrm{T}})\left(u_{x(x+\Delta x,z-\Delta z)} - u_{x(x+\Delta x,z+\Delta z)}\right)\right]. \tag{3.38a}$$

$$\tilde{u}_{z(x+\Delta x,z)} = \frac{1}{1+\varepsilon_{\mathrm{T}}}\left[u_{z(x+\Delta x,z)} + \varepsilon_{\mathrm{T}} u_{z(x,z)} - \frac{1}{4}(1+2\varepsilon_{\mathrm{T}})\left(u_{x(x,z+\Delta z)} - u_{x(x,z-\Delta z)}\right) + \frac{1}{4}\left(u_{x(x+\Delta x,z+\Delta z)} - u_{x(x+\Delta x,z-\Delta z)}\right)\right]. \tag{3.38b}$$

The compact formats for the fictitious displacements in Eqs. (3.37a, 3.37b and 3.38a, 3.38b) obtained from the nonwelded boundary conditions at the x-normal fracture interface $Z = \left(x + \frac{1}{2}\Delta x, z\right)$ are

$$\tilde{\boldsymbol{u}}_{(x,z)_{\text{nonwelded}}} = \mathrm{FNF}\boldsymbol{u}_{(x,z)} + \mathrm{FMF}\boldsymbol{u}_{(x+\Delta d,z)} - \frac{1}{4}\mathrm{FEF}\left(\boldsymbol{u}_{(x,z-\Delta d)} - \boldsymbol{u}_{(x,z+\Delta d)}\right) + \frac{1}{4}\mathrm{FGF}\left(\boldsymbol{u}_{(x+\Delta d,z-\Delta d)} - \boldsymbol{u}_{(x+\Delta d,z+\Delta d)}\right), \tag{3.39}$$

$$\tilde{\boldsymbol{u}}_{(x+\Delta d,z)_{\text{nonwelded}}} = \mathrm{FNF}\boldsymbol{u}_{(x+\Delta d,z)} + \mathrm{FMF}\boldsymbol{u}_{(x,z)} - \frac{1}{4}\mathrm{FGF}\left(\boldsymbol{u}_{(x,z+\Delta d)} - \boldsymbol{u}_{(x,z-\Delta d)}\right) + \frac{1}{4}\mathrm{FEF}\left(\boldsymbol{u}_{(x+\Delta d,z+\Delta d)} - \boldsymbol{u}_{(x+\Delta d,z-\Delta d)}\right), \tag{3.40}$$

where

$$\mathrm{F} = \begin{bmatrix} 0 & 1 \\ 1 & 0 \end{bmatrix}. \tag{3.41}$$

The nearest-neighbor fictitious displacement points $\tilde{\boldsymbol{u}}_{(x-\Delta d,z)}$ and $\tilde{\boldsymbol{u}}_{(x,z\pm\Delta d)}$ obtained from the welded boundary conditions are identical to Eqs. (3.19)–(3.21) in this section.

Note that the nonweldedness parameters $\varepsilon_{\mathrm{T}}, \delta_{\mathrm{N}}$ and γ_{N} have the same expressions as Eq. (3.14). However, they are computed from the vertical nonwelded boundary conditions. Physically, ε_{T} determines the contribution to the normal displacement (u_z) discontinuity across the fracture interface from the continuous change in the displacement, while δ_N and γ_N determine the contribution to the tangential displacement (u_x) discontinuity across the fracture interface from the continuous change in the displacement.

3.4.3 Equation of Motion

As discussed above, a P-SV wave equation of the motion can be approximated as Eq. (3.28) by taking Taylor's first order approximation (Eq. 3.27). In order to represent all boundary conditions including the medium discontinuity on the vertical fracture, the real displacement at the nearest-neighbor points should be replaced by the nearest-neighbor fictitious displacement in the approximated wave Eq. (3.28).

$$\begin{aligned} \boldsymbol{u}^{t+1}_{(x,z)} &= -\boldsymbol{u}^{t-1}_{(x,z)} + 2\boldsymbol{u}^{t}_{(x,z)} \\ &+ \left(\frac{\Delta t}{\Delta x}\right)^2 \mathrm{A}\left(\tilde{\boldsymbol{u}}^{t}_{(x+\Delta x,z)_{\text{nonwelded}}} - 2\boldsymbol{u}^{t}_{(x,z)} + \tilde{\boldsymbol{u}}^{t}_{(x-\Delta x,z)}\right) \\ &+ \left(\frac{\Delta t}{\Delta z}\right)^2 \mathrm{C}\left(\tilde{\boldsymbol{u}}^{t}_{(x,z+\Delta z)} - 2\boldsymbol{u}^{t}_{(x,z)} + \tilde{\boldsymbol{u}}^{t}_{(x,z-\Delta z)}\right). \end{aligned} \tag{3.42}$$

Substituting all evaluated fictitious displacement points into Eq. (3.42), the P-SV equation of motion for a homogeneous vertically fractured medium is formulated as

$$\begin{aligned} \boldsymbol{u}^{t+1}_{(x,z)} &= -\boldsymbol{u}^{t-1}_{(x,z)} + 2\boldsymbol{u}^{t}_{(x,z)} \\ &+ \left(\frac{\Delta t}{\Delta d}\right)^2 A\left[\left(\boldsymbol{u}^{t}_{(x+\Delta d,z)} - 2\boldsymbol{u}^{t}_{(x,z)} + \boldsymbol{u}^{t}_{(x-\Delta d,z)}\right)\right] \\ &+ \frac{1}{4}\left(\frac{\Delta t}{\Delta d}\right)^2 B\left[\boldsymbol{u}^{t}_{(x+\Delta d,z+\Delta d)} - \boldsymbol{u}^{t}_{(x-\Delta d,z+\Delta d)} - \boldsymbol{u}^{t}_{(x+\Delta d,z-\Delta d)} + \boldsymbol{u}^{t}_{(x-\Delta d,z-\Delta d)}\right] \\ &+ \left(\frac{\Delta t}{\Delta d}\right)^2 C\left[\boldsymbol{u}^{t}_{(x,z+\Delta d)} - 2\boldsymbol{u}^{t}_{(x,z)} + \boldsymbol{u}^{t}_{(x,z-\Delta d)}\right] \\ &+ \left(\frac{\Delta t}{\Delta d}\right)^2 A(\mathrm{FMF})\left[\boldsymbol{u}^{t}_{(x,z)} - \boldsymbol{u}^{t}_{(x+\Delta d,z)} - \frac{1}{4}H^{\mathrm{T}}\left(\boldsymbol{u}^{t}_{(x+\Delta d,z+\Delta d)} - \boldsymbol{u}^{t}_{(x+\Delta d,z-\Delta d)} \right.\right. \\ &\left.\left. + \boldsymbol{u}^{t}_{(x,z+\Delta d)} - \boldsymbol{u}^{t}_{(x,z-\Delta d)}\right)\right], \end{aligned} \tag{3.43}$$

where H^{T} is the transpose of H and given in Eq. (3.23). As a similarity, this equation can be used to model the sets of the vertical fractures rather than a single fracture in the fractured medium with impedance contrast.

3.5 Finite-Difference Scheme for Orthogonally Fractured Medium

3.5.1 Boundary Conditions and Fictitious Displacement Formulas

In seismology, the crust of the earth that consists of commonly layered deposits exhibits the intrinsic anisotropy of a transversely isotropic medium with a vertical symmetry axis (VTI) where the layer thickness is very much thinner than seismic wavelength. Additionally, the near vertical fractures caused by geological movements exhibit the induced anisotropy of a transversely isotropic medium with horizontal symmetric axis (HTI). Hence, it is difficult to adequately delineate the fractured reservoir using a VTI medium or a HTI medium alone. Existing geological data (e.g. outcrops) indicate that an orthorhombic model should be a more realistic model for a fractured reservoir (Tsvankin et al. 2010).

Consider a vertical slip interface that is embedded in a horizontally fractured homogeneous medium. Following the long wavelength assumption and the linear slip theory, this composed medium explicitly is equivalent to an orthogonally fractured medium that has two orthorhombic fractures at $Z = \left(x, z + \frac{1}{2}\Delta z\right)$ and $X = \left(x + \frac{1}{2}\Delta x, z\right)$. The two fractures laterally and vertically divide the composite orthorhombic homogenous medium into four spaces that are similar to the four quadrants of a plane used in mathematics. Medium 1 occupies the first quadrant with density ρ_1, S-wave velocity β_1 and P-wave velocity α_1, medium 2 occupies the second quadrant with density ρ_2, S-wave velocity β_2 and P-wave velocity α_2, medium 3 occupies the third quadrant with density ρ_3, S-wave velocity β_3 and P-wave velocity α_3, and medium 4 resides in the fourth quadrant with density ρ_4, S-wave velocity β_4 and P-wave velocity α_4 (see Table 2.1).

If the host medium of the horizontally fractured medium is an isotropic medium, the two orthogonal fracture interfaces divide the media into four identical blocks. Then, $\rho_1 = \rho_2 = \rho_3 = \rho_4; \beta_1 = \beta_2 = \beta_3 = \beta_4; \alpha_1 = \alpha_2 = \alpha_3 = \alpha_4$ (Fig. 3.4).

The orthogonally fractured medium contain three symmetric planes,$[y, z]$, $[x, z]$ and $[x, y]$, that are orthogonal to each other and normal to the $x-$, $y-$ and $z-$ axes, respectively. The evaluated central rectangle point at (x, z) should relate to the four nearest-neighbor fictitious points at $(x \pm \Delta x, z)$ and $(x, z \pm \Delta z)$, as well as the four grid boundaries at $X = \left(x \pm \frac{1}{2}\Delta x, z\right)$ and $Z = \left(x, z \pm \frac{1}{2}\Delta z\right)$. It is especially assumed that the wave propagation in this orthorhombic media will encounter two fractures as nonwelded contact boundaries at $Z = \left(x, z + \frac{1}{2}\Delta z\right), X = \left(x + \frac{1}{2}\Delta x, z\right)$ and two welded contact boundaries at $Z = \left(x, z - \frac{1}{2}\Delta z\right)$ and $X = \left(x - \frac{1}{2}\Delta x, z\right)$. Thus, the boundary conditions are successively expressed as

$$\begin{cases} u^{+}_{x(x,z+\frac{1}{2}\Delta z)} - u^{-}_{x(x,z+\frac{1}{2}\Delta z)} = S_{\mathrm{T}}\sigma^{+}_{xz(x,z+\frac{1}{2}\Delta z)} \\ \sigma^{+}_{xz(x,z+\frac{1}{2}\Delta z)} = \sigma^{-}_{xz(x,z+\frac{1}{2}\Delta z)} \end{cases}, \tag{3.44a}$$

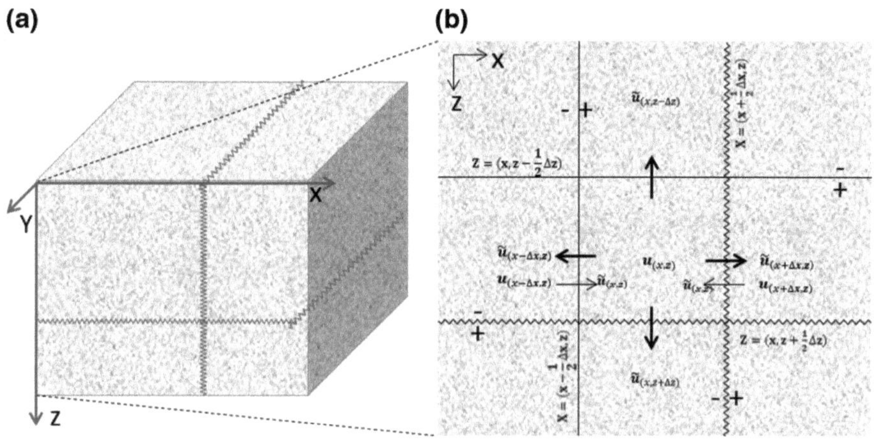

Fig. 3.4 Orthogonally fractured medium (**a**) and a finite-difference stencil (**b**) with orthorhombic fractures in the x, z-domain

$$\begin{cases} u^{+}_{z(x,z+\frac{1}{2}\Delta z)} - u^{-}_{z(x,z+\frac{1}{2}\Delta z)} = S_{\mathrm{N}}\sigma^{+}_{zz(x,z+\frac{1}{2}\Delta z)} \\ \sigma^{+}_{zz(x,z+\frac{1}{2}\Delta z)} = \sigma^{-}_{zz(x,z+\frac{1}{2}\Delta z)} \end{cases}, \tag{3.44b}$$

$$\begin{cases} u^{+}_{x(x+\frac{1}{2}\Delta x,z)} - u^{-}_{x(x+\frac{1}{2}\Delta x,z)} = S_{\mathrm{N}}\sigma^{+}_{xx(x+\frac{1}{2}\Delta x,z)} \\ \sigma^{+}_{xx(x+\frac{1}{2}\Delta x,z)} = \sigma^{-}_{xx(x+\frac{1}{2}\Delta x,z)} \end{cases}, \tag{3.45a}$$

$$\begin{cases} u^{+}_{z(x+\frac{1}{2}\Delta x,z)} - u^{-}_{z(x+\frac{1}{2}\Delta x,z)} = S_{\mathrm{T}}\sigma^{+}_{zx(x+\frac{1}{2}\Delta x,z)} \\ \sigma^{-}_{zx(x+\frac{1}{2}\Delta x,z)} = \sigma^{+}_{zx(x+\frac{1}{2}\Delta x,z)} \end{cases}, \tag{3.45b}$$

$$\begin{cases} u^{-}_{x(x,z-\frac{1}{2}\Delta z)} = u^{+}_{x(x,z-\frac{1}{2}\Delta z)} \\ \sigma^{-}_{xz(x,z-\frac{1}{2}\Delta z)} = \sigma^{+}_{xz(x,z-\frac{1}{2}\Delta z)} \end{cases}, \tag{3.46a}$$

$$\begin{cases} u^{-}_{z(x,z-\frac{1}{2}\Delta z)} = u^{+}_{z(x,z-\frac{1}{2}\Delta z)} \\ \sigma^{-}_{zz(x,z-\frac{1}{2}\Delta z)} = \sigma^{+}_{zz(x,z-\frac{1}{2}\Delta z)} \end{cases}, \tag{3.46b}$$

$$\begin{cases} u^{+}_{z(x-\frac{1}{2}\Delta x,z)} = u^{-}_{z(x-\frac{1}{2}\Delta x,z)} \\ \sigma^{+}_{zx(x-\frac{1}{2}\Delta x,z)} = \sigma^{-}_{zx(x-\frac{1}{2}\Delta x,z)} \end{cases}, \tag{3.47a}$$

$$\begin{cases} u^{+}_{x(x-\frac{1}{2}\Delta x,z)} = u^{-}_{x(x-\frac{1}{2}\Delta x,z)} \\ \sigma^{+}_{xx(x-\frac{1}{2}\Delta x,z)} = \sigma^{-}_{xx(x-\frac{1}{2}\Delta x,z)} \end{cases}, \tag{3.47b}$$

where

$$\sigma^{\pm}_{xz(x,z\pm\frac{1}{2}\Delta z)} = \mu^{\pm}_{(x,z\pm\frac{1}{2}\Delta z)}\left(\frac{\partial u^{\pm}_{x(x,z\pm\frac{1}{2}\Delta z)}}{\partial z} + \frac{\partial u^{\pm}_{z(x,z\pm\frac{1}{2}\Delta z)}}{\partial x}\right), \tag{3.48a}$$

$$\sigma^{\pm}_{zz(x,z\pm\frac{1}{2}\Delta z)} = (\lambda+2\mu)^{\pm}_{(x,z\pm\frac{1}{2}\Delta z)}\frac{\partial u^{\pm}_{z(x,z\pm\frac{1}{2}\Delta z)}}{\partial z} + \lambda^{\pm}_{(x,z\pm\frac{1}{2}\Delta z)}\frac{\partial u^{\pm}_{x(x,z\pm\frac{1}{2}\Delta z)}}{\partial x}, \tag{3.48b}$$

$$\sigma^{\pm}_{zx(x\pm\frac{1}{2}\Delta x,z)} = \mu^{\pm}_{(x\pm\frac{1}{2}\Delta x,z)}\left(\frac{\partial u^{\pm}_{z(x\pm\frac{1}{2}\Delta x,z)}}{\partial x} + \frac{\partial u^{\pm}_{x(x\pm\frac{1}{2}\Delta x,z)}}{\partial z}\right), \tag{3.49a}$$

$$\sigma^{\pm}_{xx(x\pm\frac{1}{2}\Delta x,z)} = (\lambda+2\mu)^{\pm}_{(x\pm\frac{1}{2}\Delta x,z)}\frac{\partial u^{\pm}_{x(x\pm\frac{1}{2}\Delta x,z)}}{\partial x} + \lambda^{\pm}_{(x\pm\frac{1}{2}\Delta x,z)}\frac{\partial u^{\pm}_{z(x\pm\frac{1}{2}\Delta x,z)}}{\partial z}. \tag{3.49b}$$

Formulae for the two fictitious displacements $\tilde{\boldsymbol{u}}_{(x,z+\Delta z)_{\text{nonwelded}}}$ and $\tilde{\boldsymbol{u}}_{(x+\Delta x,z)_{\text{nonwelded}}}$ obtained from the nonwelded boundary conditions are given in Eqs. (3.16) and (3.40) respectively. Solutions for the two fictitious displacements $\tilde{\boldsymbol{u}}_{(x,z-\Delta z)}$ and $\tilde{\boldsymbol{u}}_{(x-\Delta x,z)}$ are obtained from the welded boundary conditions and presented in Eqs. (3.20) and (3.21).

3.5.2 Equation of Motion

Following the horizontally and vertically fractured medium procedures, by applying the homogeneous approach for FD wave modeling, one should replace the nearest-neighbor real displacement points in Eq. (3.28) by the corresponding fictitious displacement points. In other words, the z-normal and x-normal fractures at the two nonwelded contact boundaries at $X = \left(x + \frac{1}{2}\Delta x, z\right)$ and $Z = \left(x, z + \frac{1}{2}\Delta z\right)$ have been imposed explicitly. One obtains

$$\begin{aligned}\boldsymbol{u}^{t+1}_{(x,z)} &= -\boldsymbol{u}^{t-1}_{(x,z)} + 2\boldsymbol{u}^{t}_{(x,z)}\\ &+ \left(\frac{\Delta t}{\Delta x}\right)^2 A\left(\tilde{\boldsymbol{u}}^{t}_{(x+\Delta x,z)_{\text{nonwelded}}} - 2\boldsymbol{u}^{t}_{(x,z)} + \tilde{\boldsymbol{u}}^{t}_{(x-\Delta x,z)}\right)\\ &+ \left(\frac{\Delta t}{\Delta z}\right)^2 C\left(\tilde{\boldsymbol{u}}^{t}_{(x,z+\Delta z)_{\text{nonwelded}}} - 2\boldsymbol{u}^{t}_{(x,z)} + \tilde{\boldsymbol{u}}^{t}_{(x,z-\Delta z)}\right).\end{aligned} \tag{3.50}$$

Substituting four fictitious points around the evaluated point (x, z) into Eq. (3.50) yields the P-SV equation of motion for an orthogonally fractured medium.

$$
\begin{aligned}
\boldsymbol{u}^{t+1}_{(x,z)} &= -\boldsymbol{u}^{t-1}_{(x,z)} + 2\boldsymbol{u}^{t}_{(x,z)} \\
&\quad + \left(\frac{\Delta t}{\Delta d}\right)^2 A\left[\left(\boldsymbol{u}^{t}_{(x+\Delta d,z)} - 2\boldsymbol{u}^{t}_{(x,z)} + \boldsymbol{u}^{t}_{(x-\Delta d,z)}\right)\right] \\
&\quad + \frac{1}{4}\left(\frac{\Delta t}{\Delta d}\right)^2 B\left[\boldsymbol{u}^{t}_{(x+\Delta d,z+\Delta d)} - \boldsymbol{u}^{t}_{(x-\Delta d,z+\Delta d)} - \boldsymbol{u}^{t}_{(x+\Delta d,z-\Delta d)} + \boldsymbol{u}^{t}_{(x-\Delta d,z-\Delta d)}\right] \\
&\quad + \left(\frac{\Delta t}{\Delta d}\right)^2 C\left[\boldsymbol{u}^{t}_{(x,z+\Delta d)} - 2\boldsymbol{u}^{t}_{(x,z)} + \boldsymbol{u}^{t}_{(x,z-\Delta d)}\right] \\
&\quad + \left(\frac{\Delta t}{\Delta d}\right)^2 A(\mathrm{FMF})\left[\boldsymbol{u}^{t}_{(x,z)} - \boldsymbol{u}^{t}_{(x+\Delta d,z)} - \frac{1}{4}H^{\mathrm{T}}\left(\boldsymbol{u}^{t}_{(x+\Delta d,z+\Delta d)} - \boldsymbol{u}^{t}_{(x+\Delta d,z-\Delta d)}\right.\right. \\
&\quad \left.\left. + \boldsymbol{u}^{t}_{(x,z+\Delta d)} - \boldsymbol{u}^{t}_{(x,z-\Delta d)}\right)\right] \\
&\quad + \left(\frac{\Delta t}{\Delta d}\right)^2 \mathrm{CM}\left[\boldsymbol{u}^{t}_{(x,z)} - \boldsymbol{u}^{t}_{(x,z+\Delta d)} - \frac{1}{4}H\left(\boldsymbol{u}^{t}_{(x+\Delta d,z+\Delta d)} - \boldsymbol{u}^{t}_{(x-\Delta d,z+\Delta d)}\right.\right. \\
&\quad \left.\left. + \boldsymbol{u}^{t}_{(x+\Delta d,z)} - \boldsymbol{u}^{t}_{(x-\Delta d,z)}\right)\right].
\end{aligned}
\tag{3.51}
$$

M refers to the fracture compliances and contribute to two fractured boundaries at $X = (x + \frac{1}{2}\Delta x, z)$ and $Z = (x, z + \frac{1}{2}\Delta z)$ which are dependent of the discontinuous medium. Equations (3.30), (3.43) and (3.51) express that the nonwelded contact boundary conditions govern the extra displacement discontinuity across the fracture boundary. This equation will enable us to model a complex set of the fractures if they are defined as a horizontal and vertical fracture in the medium.

3.6 Numerical Applications and Discussions

This section shows modeled wave behaviors of the horizontally, vertically and orthogonally fractured media, which gives insights into studying fracture seismic signatures.

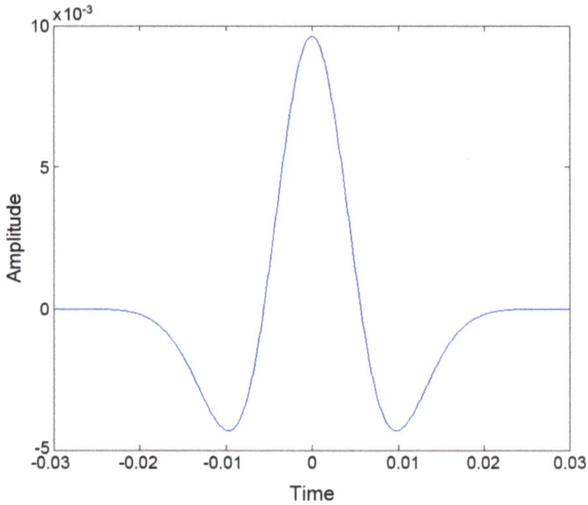

Fig. 3.5 Source wavelet. A zero-phase Ricker wavelet with dominant frequency 40 Hz

3.6.1 *Implementation of Seismic Source*

A Ricker wavelet is used as the seismic source (CREWES software)

$$\text{Ricker}(t) = \left(1 - 2\pi^2 f^2 t^2\right) \mathrm{e}^{\left(-\pi^2 f^2 t^2\right)}, \tag{3.52}$$

where "t" is time in seconds, "f" is frequency in hertz. Note that the Ricker wavelet in Eq. (3.52) is a zero-phase wavelet with a central peak and two smaller side lobes (Fig. 3.5). With a dominant frequency of $f = 40$ Hz, the breadth of the input Ricker wavelet is $0.7797/f = 0.02$ (ms). The amplitude and phase spectra of the input Ricker wavelet show that the wavelet phase is zero and the main energy is around the dominant frequency 40 Hz (Fig. 3.6). The numerical artifacts of the source have been attenuated before the wavefront impinges on the reflections once a fractured medium is modeled as the effective fractured TI medium.

3.6.2 *Stability Condition*

Using the finite-difference method requires determination of the spatial and temporal sampling intervals to meet the requirements of Courant condition, so that the numerical results can avoid the instability problem in which the wavefield grows without bound and eventually exceeds the model precision. The Courant condition for the second order n-dimensional FD for P-waves has been provided by Lines et al. (1999) as

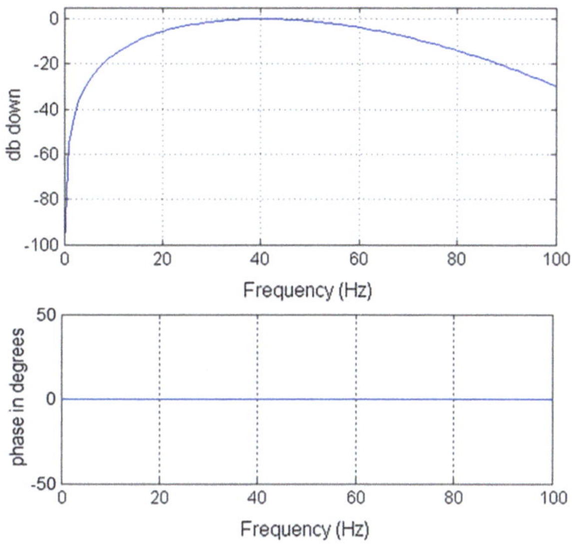

Fig. 3.6 The amplitude and phase spectrum of the source wavelet. It is clear that wavelet phase is zero and the significant energy is around frequency 40 Hz

$$a = \frac{\sqrt{n}\Delta t \alpha}{\Delta d} \approx 1, \tag{3.53}$$

where Δd is the spatial sampling interval, Δt is the time sampling interval, and the P wave velocity is α.

Manning (2008) addressed the issue of the stability condition for synthetic P-SV seismograms that was given by Kelly et al. (1976) as

$$a = \frac{\Delta t \sqrt{\alpha^2 + \beta^2}}{\Delta d} \approx 1. \tag{3.54}$$

Slawinski (1999) discussed the Courant condition for nonwelded boundary contact for the P-SV case, and showed that

$$a = \frac{\Delta t \sqrt{\alpha^2/(1+\delta_N) + \beta^2/(1+\varepsilon_T)}}{\Delta d} \approx 1, \tag{3.55}$$

where δ_N and ε_T are dimensionless nonweldedness parameters that are related the parameters of the fracture compliance discussed in Sect. 3.4. To minimize the run time and numerical dispersion, the time step is kept as a constant value with the Courant condition for the fractured medium. So

$$\Delta t = \frac{\Delta d}{\sqrt{\alpha^2/(1+\delta_N) + \beta^2/(1+\varepsilon_T)}}. \tag{3.56}$$

3.6.3 Model Parameters

The model parameters given in this section are used for three subsurface models in the book: a horizontally fractured medium, vertically fractured medium and an orthogonally fractured medium. Three fractured medium models are formed by a horizontal, a vertical and an orthorhombic fractures embedded into a homogeneous isotropic medium, respectively. The first model of the horizontally fractured medium has a P-wave velocity of $\alpha = 3790\,(\mathrm{m/s})$, a shear wave of $\beta = 2100\,(\mathrm{m/s})$, and the density is $\rho = 2.10\,(\mathrm{g/cm^3})$. The normal and tangential compliances $S_{\mathrm{N}} = 2.73 \times 10^{11}\,(\mathrm{m/Pa}), S_{\mathrm{T}} = 1.35 \times 10^{11}\,(\mathrm{m/Pa})$ can effectively induce an anisotropic properties to this horizontally fractured medium if it is needed that is equivalent to VTI medium as $\varepsilon = 0.189$ and $\delta = 0.204$ by taking Eq. (2.33) in Chap. 2. For the second model of the vertically fractured medium, the P-wave velocity is $\alpha = 3670$ (m/s), the S-wave velocity is $\beta = 2000$ (m/s), and the density $\rho = 2.41$ (g/cm^3). The normal and tangential compliances are $S_{\mathrm{N}} = 1.73 \times 10^{11}\,(\mathrm{m/Pa}), S_{\mathrm{T}} = 1.35 \times 10^{11}\,(\mathrm{m/Pa})$. The corresponding anisotropic parameters of this effective vertically fractured medium (HTI), if it is needed, are $\varepsilon^{(v)} = -0.15, \delta^{(v)} = -0.155$ that obtain from Eq. (2.37) in Chap. 2. For the third model of the orthogonally fractured medium, the P-wave velocity is $\alpha = 2900$ (m/s), the S-wave velocity is $\beta = 1600$ (m/s), and the density $\rho = 2.50$ (g/cm^3). The normal and tangential compliances are $S_{\mathrm{N}} = 3.13 \times 10^{11}\,(\mathrm{m/Pa}), S_{\mathrm{T}} = 4.76 \times 10^{11}\,(\mathrm{m/Pa})$. The resulting $S_{\mathrm{N}}, S_{\mathrm{T}}$ and dominant frequency ω should satisfy $V_p p \omega S_{\mathrm{N}} \ll 1, V_s p \omega S_{\mathrm{T}} \ll 1$ (Chaisri and Krebes 2000). We use a staggered-grid coding (Manning 2008) with second order spatial differences and second-order temporal differences. The spatial grid size Δd is 1 m and the time step Δt is 0.1 ms.

3.6.4 Horizontal Fracture Model

Figure 3.7 shows the geometry for a horizontally fractured medium model. The source is located at the center of the model (at 900 m depth). The horizontal receiver array at which the normal and tangential displacement components are recorded lies above the fracture at a distance of 155 m. The fracture itself is 1800 m long and is embedded at a depth d = 1050 m in the model. In order to investigate the fracture itself without any other reflection effects and to avoid the free surface problem and the interference with the boundary reflections, the receiver array is buried at 5 m (at 895 m depth) above the source, and the maximum horizontal offset is 900 m, and the model is a uniform homogeneous isotropic medium without anisotropic issues.

Modeled wavefield snapshots are shown in Fig. 3.8. The snapshots are the horizontal and vertical components at 0.2799 (s). The modeled wavefields contain not only direct arrivals, but also PP and PS reflected and transmitted waves which

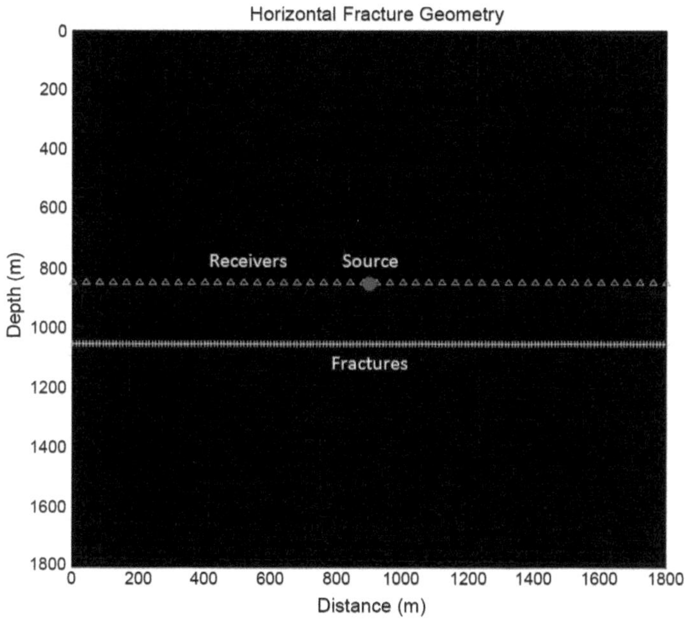

Fig. 3.7 Geometry of the horizontal fractures model of 1800 m × 1800 m. The source is located at the center of the model. The receivers are 5 m above the source. A horizontal fracture is 150 m below the source

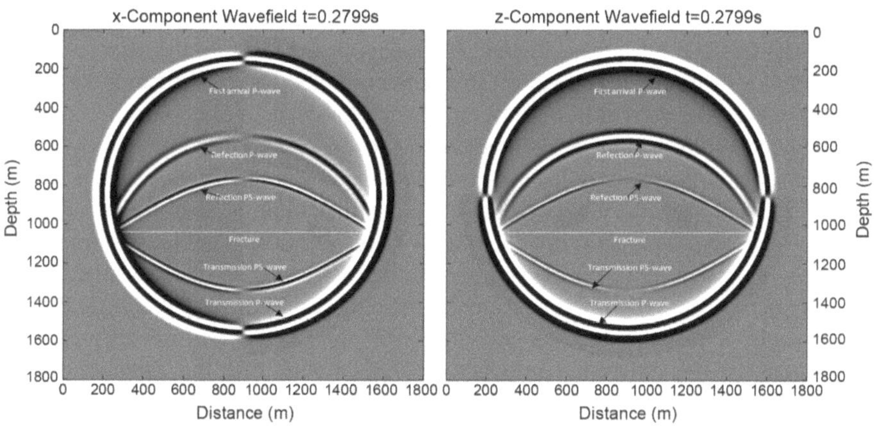

Fig. 3.8 Snapshot of wavefields. The tangential (x-component) and the normal (z-component) wavefields propagate in a uniform isotropic medium. These are not only the direct wavefields and the transmission wavefields, but also PP and PS reflection wavefields from the fracture

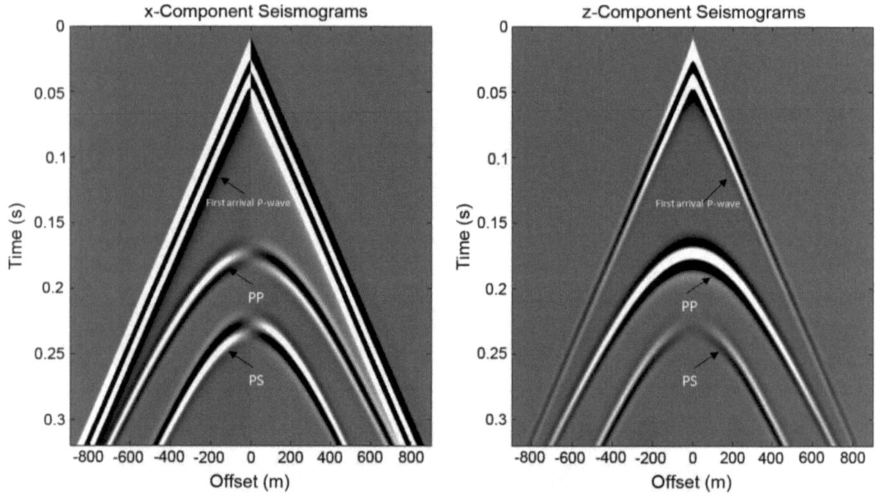

Fig. 3.9 Seismograms of the horizontal fracture. The *left side* of seismogram is *x*-components. The *right side* of the seismogram is z-component. A horizontal fracture is visible in the PP and PS reflections in the seismograms. They show that the *z*-component reflection is dominated by the PP refection, while the PS reflection amplitude dominates at the receivers in the *x*-component

are generated by the horizontal fracture. Obviously, the horizontal fracture would be detected from seismic data because the fracture is a reflector that generates the PP and PS reflected waves which are then recorded at the receivers.

Figure 3.9 shows the recorded seismograms in *x*, *z*-components simulated from a horizontal fracture subsurface model. The media parameters are the same as model parameters of the uniform medium in the Sect. 3.6.3. The left seismogram is the *x*-component. The right side seismogram is *z*-component. A phenomenon observed is that the near offset *z*-component reflections are dominated by PP waves, while the near offset *x*-component reflections are dominated by PS-waves. As the offsets increase, this phenomenon decreases. The explanation is that the near offset PP-wave polarization is normal to the horizontal fracture in the *z*-component, whereas PS-wave polarization is perpendicular to the horizontal fracture in *x*-component. Consequently, the normally incident waves on the horizontal fracture are entirely reflected, and then result in PP and PS reflections with different amplitudes in the different components. This implies that the direction of the fracture can be determined by the dominant wave amplitudes in the multicomponent data.

Figure 3.10 shows the horizontal fracture seismograms of negative offset with PP and PS amplitudes variation in the *x* and *z*-components (using CREWES software). The left graph shows the PP amplitudes as a function of offset. The right graph shows the PS amplitudes variation with offset. The red color is for the *x*-component, and the black color is for the *z*-component. The PP amplitudes of the *x* and *z*-components cross at an offset of around 400 m, while the PS amplitudes of

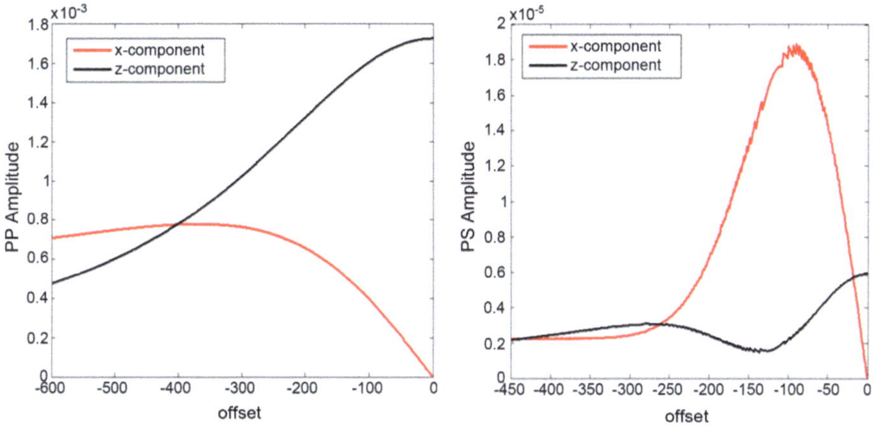

Fig. 3.10 Graph of the horizontal fracture PP and PS amplitudes, x and z-components. The *left graph* is the PP amplitudes. The *right graph* is the PS amplitudes. The *red color* is for the x-component, and the *black color* is for the z-component

the x and z-components cross around 250 m. Figure 3.10 verifies that the amplitude levels of the PP and PS waves change in the x and z-components of the seismograms in Fig. 3.9.

Figure 3.11 shows the seismic traces after AGC (Automatic gain control, CREWES software). The traces are reflections from a horizontal fracture interface (black) and a horizontal small impedance contrast interface (red) at a near offset of 5 m and a far offset of 420 m. The media parameters are the same as the ones in Sect. 3.6.3. The first arrivals are the same since they are not distorted by any reflection or transmission at all. However, both reflection coefficients are different and have a phase rotation.

Figure 3.12 demonstrates snapshots of the wavefields in the horizontally fractured medium in x-component (the left) and z-component (the right). The top row shows the modeling of the wave propagation in the horizontally fractured medium that is irrespective anisotropic problem, while the bottom row illustrates the wave propagation in the same medium as the top one, but the fractured medium is regarded as a fracture-induced anisotropic medium. Figure 3.12 exhibits that the wavefronts is a perfect circle in the top row, whereas the wavefronts making an ellipse shape with a lateral longer axis in the bottom row means that the traveltime is angularly variant. This manifests that the vertical wavefront is less advanced than horizontal wavefront when wave propagate in the effective horizontally fractured anisotropic medium (VTI medium). This conclusion agrees to an opinion about different incident angle with different traveltime in a VTI anisotropic medium (Rueger 2002).

Figure 3.13 illustrates the reflections and their calculations for the media composition and decomposition that shows traces from different reflectors. The left side of the figure is for the z-component of the seismograms, and the right side of the

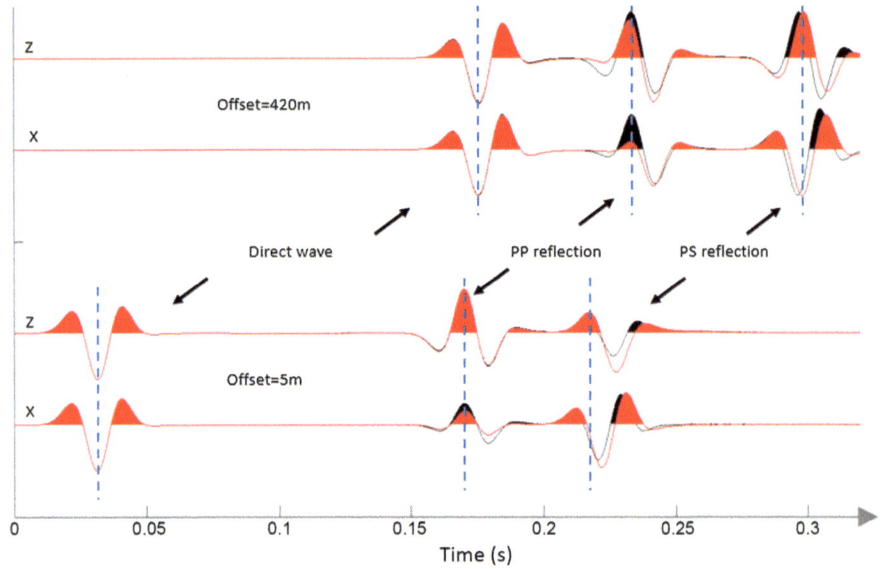

Fig. 3.11 Seismic traces after AGC (Automatic gain control). The traces are from a horizontal fracture (*black*) and a horizontal impedance contrast interface (*red*) at a near offset of 5 m and a far offset of 420 m

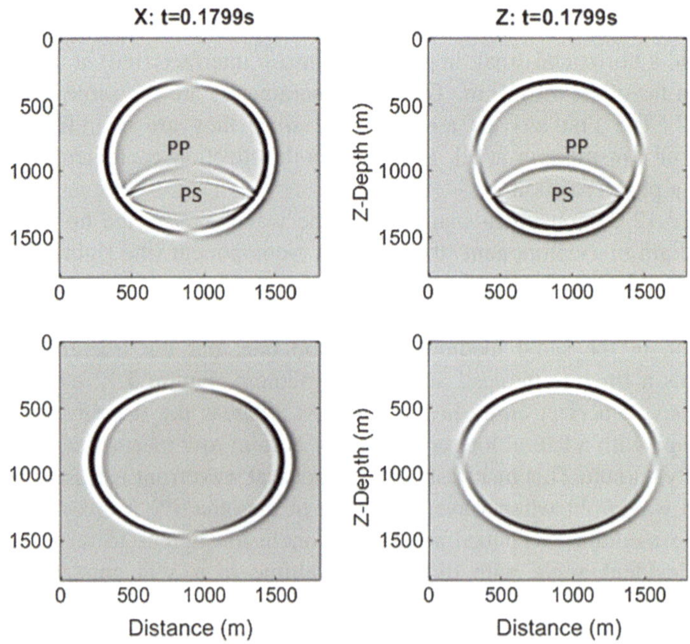

Fig. 3.12 Snapshot of wavefields in the horizontally fractured medium. The *top row* shows the wave propagation in the horizontally fractured medium with irrespective anisotropic problem. The *bottom row* illustrates the wave propagation in the same medium as *top row*, but the medium is regarded as an effective horizontally fractured anisotropic medium

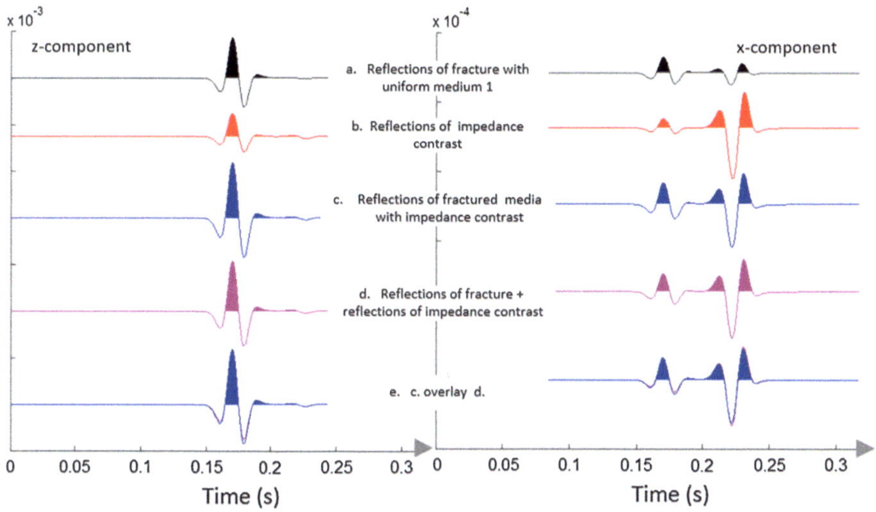

Fig. 3.13 Diagram of the seismograms for Schoenberg-Muir calculus theory. The *left side* are reflections for the *z*-component. The *right side* are reflections for the *x*-component. The reflection of fractured medium with impedance contrast ↔ reflection of the fracture with uniform medium + reflection of the host media with impedance contrast

figure is for the *x*-component of the seismograms. In Fig. 3.13, the black trace (a) records reflections from the horizontally fractured medium with uniform medium 1, the red trace (b) records reflections from the impedance contrast interface, the blue trace (c) is for reflections from the horizontally fractured medium with impedance contrast, the magenta trace (d) is the summation of the first two traces (a) and (b), and the bottom traces (e) are overlain displays of the summation reflection trace (magenta) and the reflection trace (blue) of the fractured medium with impedance contrast, i.e., they are superimposed. It is not difficult to conclude that at certain media parameters, the reflections of the horizontally fractured medium with impedance contrast are numerically and approximately equal to a linear summation of the reflections of the fracture medium with uniform medium 1 and the reflections of the host media with impedance contrast interface. This means that the reflections of the horizontally fractured medium with impedance contrast numerically can be approximately separated into the reflections of the fracture and the reflections of the unfractured host media, i.e., at certain media parameters, reflection from fractured medium with impedance contras ≈ reflection from fracture with uniform medium 1 + reflection from host medium with impedance contrast. This investigation is for a preparation of the fractured medium inversion in the next chapter.

3.6.5 Vertical Fracture Model

Figure 3.14 shows the vertically fractured medium model. The source is located at the center of the model and the vertical fracture is 50 m away at right positive offset side. The receiver arrays is arranged horizontally above the source at a distance of 100 m. The other model parameters are the same as those of the horizontally fractured medium model in Sect. 3.6.3.

According to Snell's law and the model geometry in Fig. 3.14, the location of the vertical fracture decides the path of the reflection wave. The relatively far offsets will receive reflections with small incident angles, while the relatively near offsets will record the waves having larger incident angles. This is different from the horizontally fractured medium case in which relatively far offsets receive waves with large incident angles, while relatively near offsets record waves with small incident angles. A cartoon in Fig. 3.15 illustrates the path of reflection waves in the wavefield for this vertically fractured medium.

The vertical fracture is detectable and visible through PP and PS reflections in the seismograms (Fig. 3.16) as well. In Fig. 3.16, the direct wave has been removed

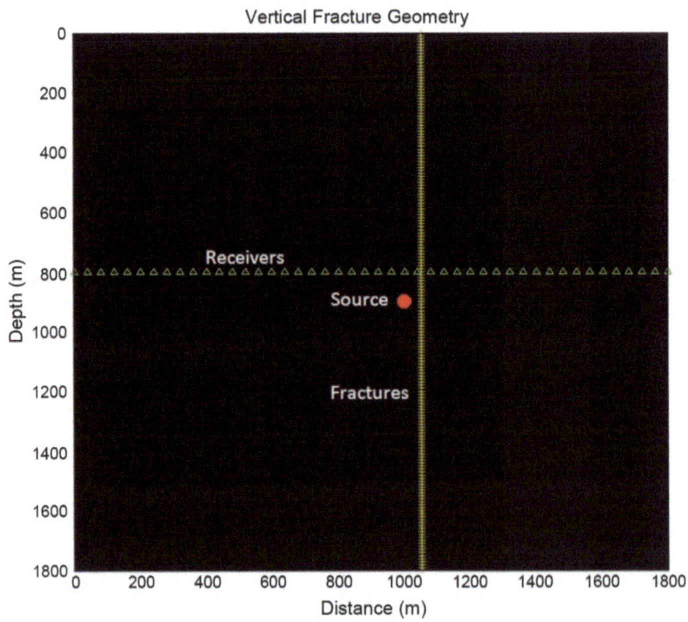

Fig. 3.14 Geometry of the vertical fractures model of 1800 m × 1800 m. The source is located in central of the model. The receivers are 100 m above the source. A vertical fracture is located 50 m to the right of the source

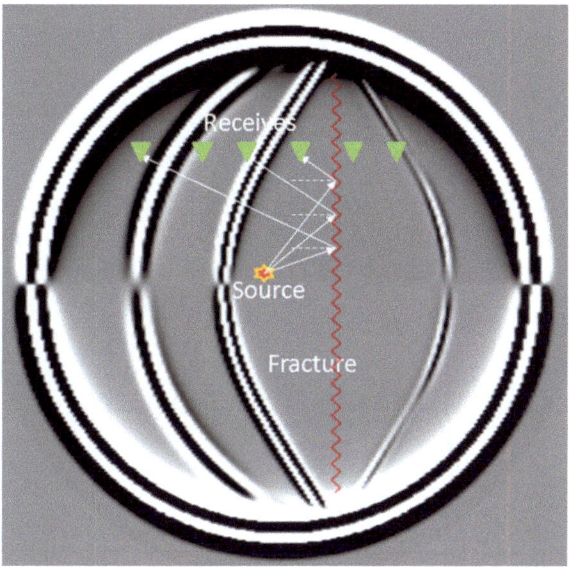

Fig. 3.15 Picture of the path of reflection and the wavefields of the vertically fractured medium. The relatively far offsets will receive reflections with small incident angles, while the relatively near offsets will record the waves with the larger incident angles

from the original records, and the negative offset traces record the reflection waves, while the positive offset traces record the transmission waves. The left side shows the seismograms for the x-component, and the right side shows the seismograms for the z-component. Note that the amplitude differences between the PP and PS reflections are opposite to those of the case of the horizontally fractured medium, because the direction of the fracture has been changed. The amplitude of the PP wave dominates in the x-component, while the PS amplitude dominates in the z-component. This further proves that PP and PS amplitudes are related to the direction of the existing fracture. In other words, the change in the amplitude levels of elastic waves in multicomponent data can be used to detect the direction of the existing fracture. It denotes that the PS reflection in the z-component propagating in the vertical fracture plane coincides with a fast wave of the split PS data of the multicomponent seismic data.

Figure 3.17 shows snapshots of the wavefields in a vertically fractured effective anisotropic HTI medium. The left side and right side of Fig. 3.17 are for x, z-components, respectively. Obviously, in both components, the horizontal wavefront is less advanced than the vertical wavefront that agrees with a conclusion of the traveltime (amplitude) azimuth-dependent anisotropic properties for the HTI medium (Rueger 2002).

The reason is that the vertical wavefront is the wave propagation closing to the vertical fracture plane or an isotropic plane, whereas the horizontal wavefront presents the direction of wave propagation that is normal to the fracture plane. In

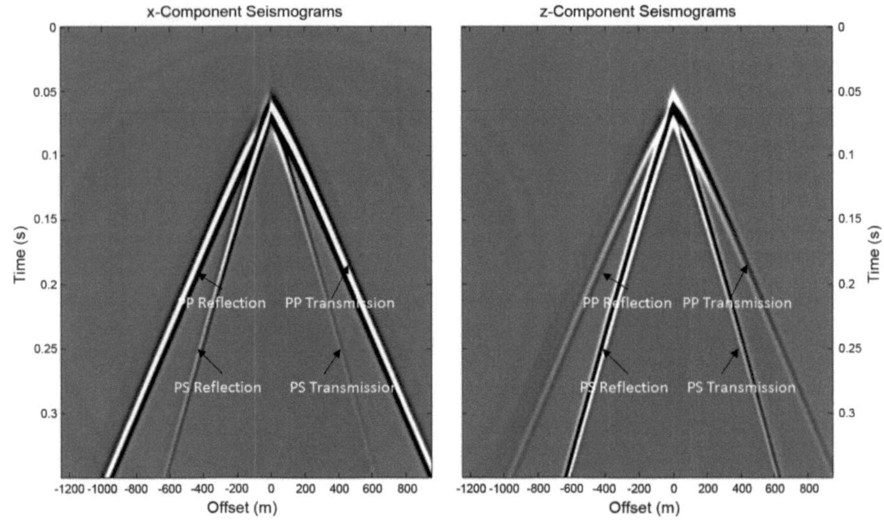

Fig. 3.16 Seismograms of the vertical fracture. The *left side* is the *x*-component and the *right side* is the *z*-component. The vertical fracture is visible and detectable through the PP and PS reflections in the seismograms. The amplitudes of the PP wave dominate in the *x*-component, while the amplitudes of the PS wave dominate the *z*-component

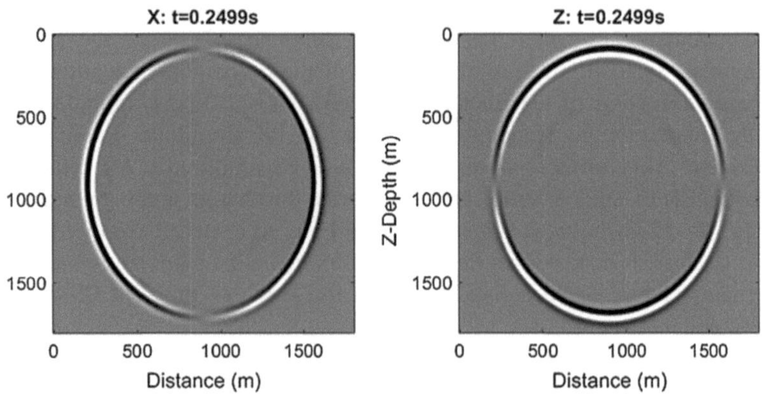

Fig. 3.17 Snapshots of the wavefields in a vertically fractured effective anisotropic HTI medium. The *left side* and *right side* are for *x*, *z*-components, respectively

particular, for this geometry of the vertically fractured medium, each receiver and the source make different planes that have different azimuths relative to the fracture plane (Fig. 3.18). This vertically fractured model fully presents HTI characteristics

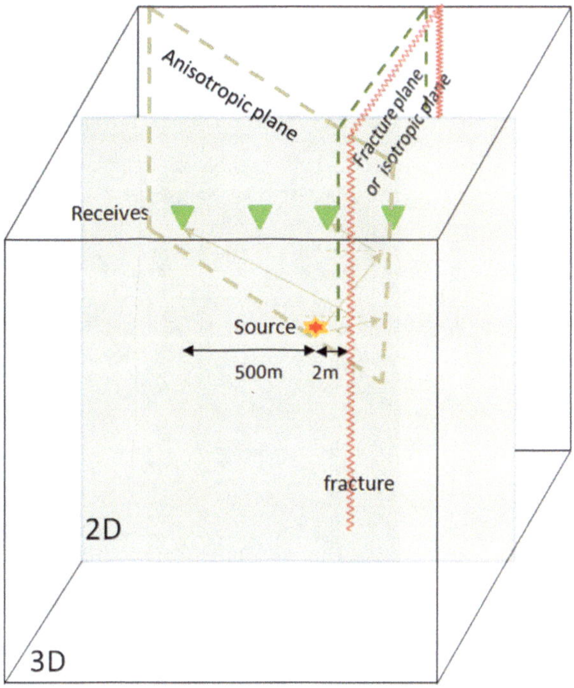

Fig. 3.18 Schematic of fracture with azimuth issues. Each receiver and the source make planes with different azimuths with respect to the fracture plane. The receivers with near offsets and the source make planes that are closely parallel to fracture plane, while the receivers with far offsets and the source create planes that are nearly normal to the fracture plane

in that wave amplitude and traveltime are actually azimuth-dependent. Therefore, the data from a mature fractured reservoir necessarily need anisotropic correction and azimuthal AVO analysis.

3.6.6 Orthorhombic Fracture Model

As we know, the most realistic medium model in the oil & gas reservoir is the orthorhombic medium model that can be simply simulated by integrating the models of horizontally fractured media with a vertical fracture. Figure 3.19 shows the geometry of the model of the orthorhombic medium of 1801 m × 1801 m. The source is located at the center of the model. The horizontal receiver array lies 5 m above the source. The fractures are vertically and horizontally 150 m away from the source.

Figure 3.20 shows the seismograms for the x and z-components (the top) and the x and z-component snapshots (the bottom) of the wavefield at t = 0.319 (s) of the orthorhombic fractures model. Note that the recorded synthetic seismic data and

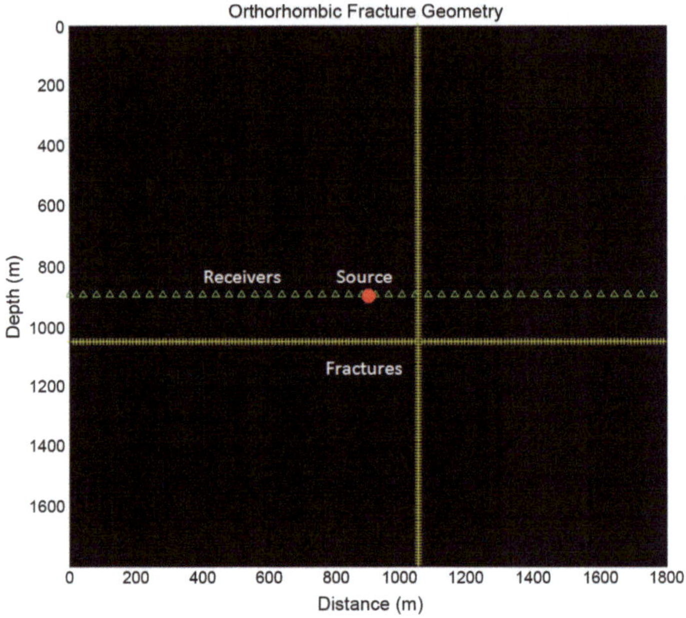

Fig. 3.19 Geometry model of the orthorhombic medium (1801 m × 1801 m). The source is located at the center of the model. The horizontal receiver array lies 5 m above the source. The fractures are 150 m vertically and horizontally away from the source

wavefield are more complex than those of the single fracture model. And all conclusions about the limited horizontally and vertically fractured media remain valid for the amplitude level in this orthorhombic fractures model. The PS-wave reflected from the horizontal fracture (PS_h) and PP-wave generated from the vertical fracture (PP_v) dominate in the *x*-component. The *z*-component seismogram is dominated by the PS-wave and PP-wave from the vertical fracture and horizontal fracture (PS_v and PP_h), respectively. The waves are reflected or converted a second time near the fracture intersection point, and these are annotated in the snapshots of the *x* and *z*-components of the wavefield as $PP_hP_v, PP_hS_v, PP_vP_h$ and PP_vS_h. The second reflected wave and the converted waves are PP_hP_v and PP_hS_v, and denote that the PP wave is reflected from the horizontal fracture first and then is reflected and converted to a P wave and S wave respectively at the vertical fracture again, while the reflected and converted waves, PP_vP_h and PP_vS_h, indicate that the PP wave is reflected from the vertical fracture first and then is reflected and converted to a P wave and S wave respectively by the vertical fracture a second time.

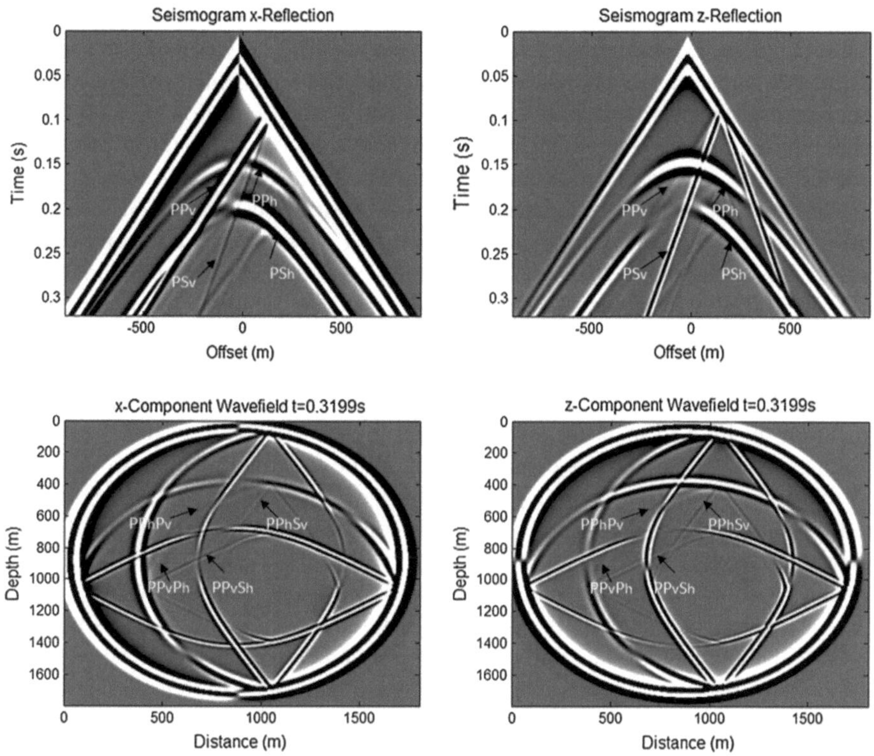

Fig. 3.20 Seismograms of orthorhombic fractures. The *top* illustrates the seismograms for the x and z components. The *bottom* shows the corresponding snapshots of the wavefield at t = 0.3199 (s). The amplitude of the PS-wave from the horizontal fracture ($\mathbf{PS_h}$) and the PP-wave from the vertical fracture ($\mathbf{PP_v}$) dominate in the x-component. The amplitude in the z-component seismograms is dominated by the PS-wave and the PP-wave from the vertical fracture and horizontal fracture ($\mathbf{PS_v}$ and $\mathbf{PP_h}$), respectively. The waves are reflected or converted a second time near the fracture intersection point and are denoted in the snapshots of the wavefield as, $\mathbf{PP_hP_v}$, $\mathbf{PP_hS_v}$, $\mathbf{PP_vP_h}$ and $\mathbf{PP_vS_h}$

3.7 Conclusions

Forward modeling of the fracture provides a connection from the subsurface models to the surface seismograms that is key to understanding geological body representations including the fracture features in the seismic data.

In this chapter, the homogeneous formulation approach of finite-difference modeling has been selected. The advantage of the homogeneous formulation approach of FD modeling for the discontinuity medium is that the boundary conditions can be implemented explicitly. This approach is suitable for modeling fractured media, for which the discontinuous displacement nonwelded boundary conditions need to be explicitly implemented.

The concept of the fictitious grid point has been studied and it has been found that it facilitates expressing the relationships between the stresses and displacements at the two sides of the boundary. Additionally, the replacement of the real displacement with the fictitious displacement in the wave equation of motion adequately presents the wave propagation on the two sides of the fractured medium, so that the waves are imprinted with the fracture properties in the seismogram data.

Theoretically, a finite-difference stencil has been discussed by considering the different boundary conditions. It is based on using only one grid boundary of a FD cell to satisfy the nonwelded boundary conditions, instead of all four boundaries (Slawinski 1999) for model the fractures. As well, the fictitious displacement formula that relates a finite-different stencil with welded and nonwelded boundary conditions has been derived. Therefore, the finite-difference schemes for a horizontally fractured medium, a vertically fractured medium and an orthogonally fractured medium have been developed and presented. The new finite-difference schemes treat the case of waves propagating in the fractured medium in which the normal equation of wave motion governs the wave propagation in the host medium and the nonwelded boundary conditions constrain the waves at the fracture interface.

The new finite-difference schemes were implemented using MATLAB code to generate the seismograms for the different fractured media. Firstly, the seismograms indicate that the first arrivals for both the fracture and impedance contrast interfaces are the same because they are not distorted by any reflection or transmission at all if the medium properties are the same. Fractures are good reflectors because they produce not only the transmitted PP and converted PS waves, but also reflected PP and PS waves with x and z components in the seismograms from a uniform homogenous isotropic medium without an impedance contrast. Conventionally, there are no reflected waves in the uniform homogenous isotropic medium if there is no fracture existing. Thus this phenomenon can be used to detect fractures in uniform lithology zones in the reservoir such as fractures in shale or coalbed zones. Secondly, the PP and PS amplitudes in the seismograms illustrate that the variation of amplitudes follows the direction of the fractures. For a horizontally fractured medium, the amplitude of the PP wave dominates in the z-component data, while the PS amplitudes are stronger than the PP amplitude in the x-component data. Oppositely, in the seismograms for the vertically fractured medium, the PP amplitudes are stronger than the PS amplitudes in the x-component, while PS amplitudes dominate in the z-component of the seismogram. Thus, the amplitudes of PP and PS waves in the x and z components of the seismograms can be used to detect the direction of the fracture. Thirdly, the reflection of the fracture has a phase rotation with respect to the reflection from the impedance contrast interface. The wavefront variation is angularly dependent so that the vertical wavefront is less advanced than the horizontal wavefront in the horizontally fractured medium. This means that the horizontal fractured medium displays VTI anisotropic properties. In the vertically fractured medium, however, the vertical wavefront close to the fracture plane is more advanced than the horizontal wavefront of the wave propagation normal to the fracture plane. This is evidence that the vertically fractured medium is equivalent to the HTI anisotropy problem: the amplitude variation is azimuth-dependent (AVAZ), i.e., different azimuths have different traveltimes (because of the different velocities).

One may conclude that the fractures are detectable and visible because fractures strongly influence the seismic wave propagation and raise issues of scattering and induced anisotropy. The amplitudes of the multicomponent seismic data are evidence that reveal the direction of the fracture. The seismic data of the fractured medium presenting anisotropy show that analyzing the anisotropy of the seismic data can provide a way to predict fractures.

References

Aki, K., & Richards, P. G. (1980). Quantitative seismology: Theory and methods (vol. 1). W.H. Freeman and Company.

Chaisri, S., & Krebes, E. S. (2000). Exact and approximate formulas for P-SV reflection and transmission coefficients for a nonwelded contact interface. *Journal Geophysical Research, 105,* 28045–28054.

Coates, R. T., & M. Schoenberg. (1995). Finite-difference modeling of faults and fractures. *Geophysics, 60,* 1514–1526.

Kelly, K. R., Ward, R. W., Treitel, S., & Alford, R. M. 1976. Synthetic seismograms finite-difference approach. *Geophysics, 41*(1), 2–27.

Korn, M., & Stöckl, H. (1982). Reflection and transmission of love channel wave at coal seam discontinuities computed with a finite difference method. *Journal of Geophysics, 50,* 171–176.

Lines, L. R., Slawinski, R., & Bording, R. P. 1999. Short note—A recipe for stability of finite-difference wave-equation computations. *Geophysics, 64*(3), 967–969.

Lines, L. R., & Newrick, R. T. (2004). *Fundamentals of geophysical interpretation.* Tulsa, Oklahoma: Society of Exploration Geophysicists Publication.

Manning, P. M. (2008). *Techniques to enhance the accuracy and efficiency of finite difference modelling for the propagation of elastic waves.* Ph.D. thesis, The University of Calgary.

Rueger, A. (2002). *Reflection coefficients and azimuthal AVO Analysis in anisotropic media: SEG geophysical monograph series number10: Soc.* Geophys: Expl.

Slawinski, R. A., & Krebes, E. S. (2002a). Finite-difference modeling of SH-wave propagation in nonwelded contact media. *Geophysics, 67,* 1656–1663.

Slawinski, R. A., & Krebes, E. S. (2002b). The homogeneous finite difference formulation of the P-SV wave equation of motion. *Studia Geophysica et Geodaetica* (Impact Factor: 0.75). 01/2002; *46*(4), 731–751. doi:10.1023/A:1021133606779

Slawinski, R. A. (1999). *Finite-difference modeling of seismic wave propagation in fractured media.* Ph.D. thesis, The University of Calgary.

Tsvankin, I., Gaiser, J., Grechka, V., van der Baan, M., & Thomsen, L. (2010). Seismic anisotropy in exploration and reservoir characterization: An overview. *Geophysics, 75,* A15–A29.

Chapter 4
Fractured Medium AVO Inversion

Abstract The synthetic seismograms for the fractures in Chap. 3 suggest that the fractures are also reflection generators for reflecting waves propagating in the media. Consequently, the recorded surface seismic data inevitably contain fracture reflections when the acquisition reservoir contains fractures. In this chapter, we derived new PP- and PS-wave reflection and transmission coefficient equations based on the principle of nonwelded boundary conditions. The new AVO equations for the horizontally fractured medium with impedance contrast are finally presented. In the new AVO equations, the reflection coefficients are divided into a welded contact part (a conventional impedance contrast part) and a nonwelded contact part (the fracture part) which makes them flexible for truly inverting for the rock properties of the fractured medium with impedance contrast.

Keywords Zoeppritz equation · Reflection/transmission coefficients · Approximate AVO inversion · Amplitudes · Seismic data processing · Horizontally fractured medium · Vertically fractured medium · Algorithm of the AVO inversion

4.1 Approximations of the Zoeppritz Equations

The Zoeppritz equations (Zoeppritz 1919) describe the relation of the incident angle of a plane *P*-wave, to the reflection and transmission coefficients of compressional waves (*P*-waves) and shear waves (*S*-waves) at an interface in a perfectly welded contact medium. The relation is the basis of the AVO and AVAZ (amplitude variation with angle and azimuth) methods which assume that the amplitudes contain impedance contrast information. The Zoeppritz equations have exact solutions but they have quite a complex form (Aki and Richards 1980). With assumptions of small fractional perturbations in elastic parameters (all of the reflectivity $\ll 1$ at zero-offset), i.e., low impedance contrast and perfectly welded contact media, many people simplified the original Zoeppritz equations into linear approximate AVO equations with different variables (Table 4.1). In 1980, Aki and

© Springer Nature Singapore Pte Ltd. 2018

X. Cui et al., *Seismic Forward Modeling of Fractures and Fractured Medium Inversion*, Springer Geophysics, DOI 10.1007/978-981-10-3584-5_4

Table 4.1 Conventional AVO equations of simplified Zoeppritz equations

Authors	AVO equations	Approximations
Aki and Richards (1980)	$R(i) \approx \frac{1}{2}(1+\tan^2\theta)\frac{\Delta\alpha}{\alpha} - \left(4\left(\frac{\beta}{\alpha}\right)^2 \sin^2\theta\right)\frac{\Delta\beta}{\beta} + \frac{1}{2}\left(1-4\left(\frac{\beta}{\alpha}\right)^2\sin^2\theta\right)\frac{\Delta\rho}{\rho}$	$\Delta\theta = \theta_2 - \theta_1 = \tan\theta\frac{\Delta\alpha}{\alpha}$, $\Delta\vartheta = \vartheta_2 - \vartheta_1 = \tan\vartheta\frac{\Delta\beta}{\beta}$. $\alpha = \frac{\alpha_2+\alpha_1}{2}$, $\Delta\alpha = \alpha_2 - \alpha_1$. $\beta = \frac{\beta_2+\beta_1}{2}$, $\Delta\beta = \beta_2 - \beta_1$. $\rho = \frac{\rho_2+\rho_1}{2}$, $\Delta\rho = \rho_2 - \rho_1$. where Δ denotes the difference in the elastic properties across the interface
Shuey (1985)	$R(\theta) \approx A + B\sin^2\theta + C\tan^2\theta\sin^2\theta$	$A = \frac{1}{2}\left(\frac{\Delta\alpha}{\alpha} + \frac{\Delta\rho}{\rho}\right)$. $B = B_0 A - \frac{\Delta v}{1-v^2}$. $B_0 = D - 2(1+D)\frac{1-2v}{1-v}$ $D = \frac{\Delta\alpha/\alpha}{\Delta\alpha/\alpha + \Delta\rho/\rho}$. $C = \frac{1}{2}\frac{\Delta\alpha}{\alpha}$ where Δv is change and v is the average Poisson ratio across the interface
Fatti et al. (1994)	$R(\theta) \approx \frac{1}{2}(1+\tan^2\theta)\frac{\Delta I_p}{I_p} - 4\left(\frac{\beta}{\alpha}\right)^2\sin^2\theta\frac{\Delta I_s}{I_s}$	$\frac{\Delta I_p}{I_p} = \frac{\Delta(\rho\alpha)}{\rho\alpha}$, $\frac{\Delta I_s}{I_p} = \frac{\Delta(\rho\beta)}{\rho\beta}$
Ruger (2002)	$R(\theta,\varphi) \approx A + (B_{\text{iso}} + B_{\text{aniso}})\sin^2\theta + (C_{\text{iso}} + C_{\text{aniso}})\tan^2\theta\sin^2\theta$	$B_{\text{iso}} = \frac{1}{2}\left(\frac{\Delta\alpha}{\alpha} - \left(\frac{2\beta}{\alpha}\right)^2\frac{\Delta G}{G}\right)$, $B_{\text{aniso}} = \frac{1}{2}\left(\Delta\delta^{(v)} + 2\left(\frac{2\beta}{\alpha}\right)^2\Delta\gamma\right)\cos^2\varphi$, $C_{\text{iso}} = \frac{1}{2}\frac{\Delta\alpha}{\alpha}$ $C_{\text{aniso}} = \frac{1}{2}\left(\Delta\delta^{(v)}\sin^2\varphi\cos^2\varphi + \Delta\varepsilon^{(v)}\cos^4\varphi\right)$ where $G = \rho\beta^2$. $\delta^{(v)}$, $\varepsilon^{(v)}$ and γ are Thomsen anisotropy parameters for HTI media
Downton et al. (2011)	$R(\theta,\varphi) = r_0(\theta) + r_2(\theta)\cos(2(\varphi-\varphi_{\text{sym}})) + r_4(\theta)\cos(4(\varphi-\varphi_{\text{sym}}))$ $r_0(\theta) = A + B\sin^2\theta + C\tan^2\theta\sin^2\theta + (C_{\text{iso}} + C_{\text{aniso}})\tan^2\theta\sin^2\theta$ $r_2(\theta) = \frac{1}{2}(B_{\text{ani}} + g(g-1)\Delta_{\text{N}}\tan^2\theta)\sin^2\theta$ $r_4(\theta) = \frac{1}{8}\kappa\sin^2\theta\tan^2\theta$	$A = A_{\text{iso}} + A_{\text{ani}} = A_{\text{iso}} - \frac{1}{4}\chi^2\Delta_{\text{N}}$ $B = B_{\text{iso}} + B_{\text{ani}} = B_{\text{iso}} + \frac{g}{2}\Delta_{\text{T}} - \frac{\chi}{4}\Delta_{\text{N}}$ $C = C_{\text{iso}} + C_{\text{ani}} = C_{\text{iso}} - \frac{1}{8}g\Delta_{\text{T}} - \frac{1}{8}(3g^2 - 4g + 2)\Delta_{\text{N}}$ where $\kappa = g(\Delta_{\text{T}} - g\Delta_{\text{N}})$, $\chi = 1 - 2g$, $g = \frac{\beta^2}{\alpha^2}$, Δ_{N}, Δ_{T} are normal and tangential weakness of the fracture

Richards presented the AVO equations by providing the elastic reflectivities of the P-velocity, S-velocity, and density. Shuey (1985) specified the AVO equation in terms of zero-offset P-wave reflectivity and a gradient. Smith and Gidlow (1987) rearranged the Aki and Richards's equation and applied an empirical relationship (Gardner et al. 1974) to the approximate AVO equations for P-wave and S-wave velocity reflectivities. Fatti et al. (1994) gave the AVO equations for P-wave and S-wave reflectivities. In 2002, Ruger used the approach of Shuey (1985) to unravel an AVAZ equation that shows the reflection coefficients are directly affected by both incident angles and azimuths. Downton et al. (2011) contributed to the AVAZ equation relation to the fractured weakness parameters through using azimuthal Fourier coefficients. Exact solutions (which are extremely complex) for the reflection and transmission coefficients for nonwelded contact boundary conditions have also been derived (Chaisri and Krebes 2000; Chaisri 2002).

4.2 Data Preconditioning for AVO/AVAZ Inversion

AVO inversion is intended to provide additional reservoir properties based on the amplitude variation over a range of offsets from the surface seismic data. Therefore, the seismic data must accurately preserve the true amplitudes in correspondence to geological factors, rather than maintain signs of non-geological bodies and artifacts from data acquisition.

4.2.1 Deconvolution

In seismology, a seismic data trace $S(t)$ is usually modeled to be the convolution of a source wavelet $W(t)$ and an earth reflectivity function $R(t)$:

$$S(t) = W(t) * R(t). \tag{4.1}$$

Seismologists are interested in the earth reflectivity $R(t)$ that represents rock formations as geological reflectors. The well-known method of deconvolution attempts to undo the convolution by applying an inverse filter $f(t)$ to remove the effects of the source wavelet $W(t)$ from the seismic data $S(t)$ in order to get back to the true reflectivity series $R(t)$ (Lines and Newrick 2004):

$$R(t) = f(t) * S(t). \tag{4.2}$$

Ideally, deconvolution (Yilmaz 1991; Margrave 2006) should provide the perfect resolution: the reflectivity function should be a series of delta functions. However, the performance of deconvolution often is imperfect because of the impacts of noisy signals, the band-limited nature of the seismic data, and the lack of knowledge about the wavelet (Lines and Newrick 2004). Nevertheless, deconvolution still compresses the seismic wavelet to achieve a flat, or white amplitude spectrum that significantly improves the resolution of seismic data. Figure 4.1 shows a real seismic data stack section and its amplitude spectrum in a certain window before and after deconvolution. The data belong to the Suping Peng research group of the State Key Laboratory of the CUMTB (China University of Mining and Technology, Beijing) and were processed by the Arcis Seismic Solution company. There is no doubt that the amplitude spectrum (orange) of the deconvolution data approaches a flat level, or the white color, meaning that the data have been enhanced with content at higher and lower frequencies. Therefore, the deconvolution is a vital step of data preconditioning.

4.2.2 Noise Attenuation

Seismic data are always contaminated by some noises that distort the true amplitudes and affect the accuracy of the AVO/AVAZ inversion results. The noises can be categorized into two types: coherent noises and random noises. The coherence noises, such as the ground roll and the multiples, can be suppressed by a Radon transform or an FK filter. For the random noises, they can be investigated in the spatial and temporal directions, and uncorrelated noise from trace to trace. Thus, a time-variant method and spatial prediction filtering such as FX deconvolution can be used to attenuate most of the random noise. Of course, conventional CMP stacking significantly reduces the uncorrelated random noises within the data. Figure 4.2 shows the result of random noise attenuation on the same real seismic data as shown in Fig. 4.1. Figure 4.2a, b shows before and after random noise

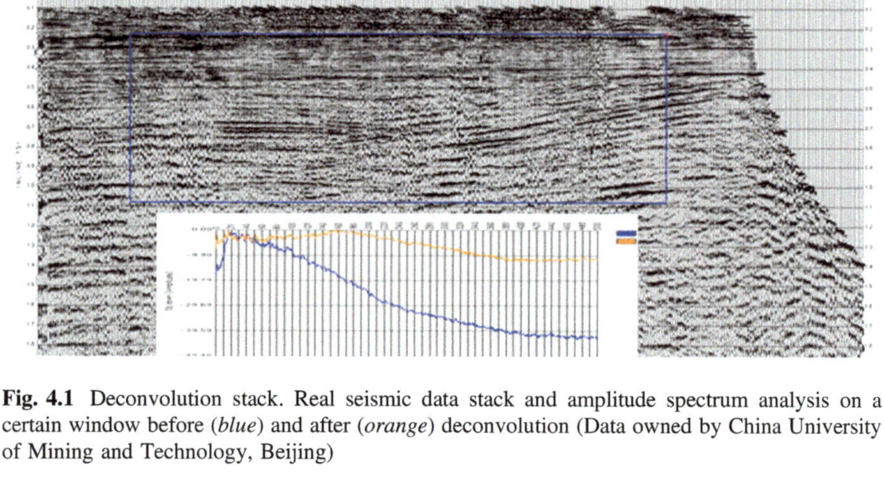

Fig. 4.1 Deconvolution stack. Real seismic data stack and amplitude spectrum analysis on a certain window before (*blue*) and after (*orange*) deconvolution (Data owned by China University of Mining and Technology, Beijing)

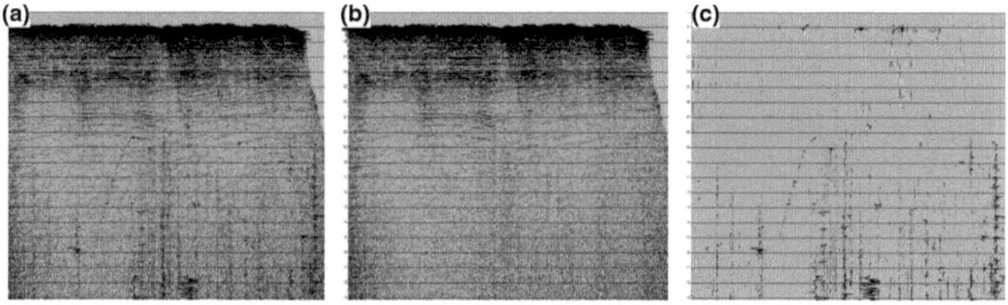

Fig. 4.2 Random noise attenuation. **a** and **b** are before and after random noise attenuation stacks, respectively. **c** is the difference between before and after noise attenuation (Data owned by China University of Mining and Technology, Beijing)

attenuation, respectively. Figure 4.2c shows the difference between before and after noise attenuation which shows no coherent reflections, which means that the random noises are attenuated and signals are preserved.

Despite more and more noise attenuation technologies having been developed, geophysicists still would like to see a lateral integration of existing technologies to enhance the ratio of signal and noise (the S/N ratio). An amplitude-friendly filtering technique, LIFT, which works for different types of noise attenuation, was published by Choo and Sudhakar (2003). They take a new approach by adding back an estimation of the signal lost during the modeling, rather than simply outputting the signal model or the signal model with a percentage of the original data added back. This approach is a practical and robust amplitude-preserving way to precondition data for the AVO process (Choo et al. 2004).

4.2.3 5D Interpolation

The purpose of the seismic data preconditioning is to enhance the signal-to-noise ratio by removing all sorts of noise. However, missing data at some offsets and azimuths can create "acquisition footprints" that are undesirable artifacts creating amplitude variations within the seismic data. In other words, the missing data or "acquisition footprints" almost always negatively impact AVO and AVAZ analyses. A method of 5D Interpolation based on Fourier reconstruction addresses the problems, in which interpolation is simultaneously performed in pre-stack data in five dimensions to predict new sources and receivers at desired locations to compensate for the missing data. The five dimensional Fourier spectra of the acquired data along inline, crossline, offset, azimuth, and time (frequency) act as a constraint on the nature of the missing traces. New interpolated traces have multidimensional Fourier spectra that are consistent with the input data in all five dimensions. Therefore, it can truly capture inline–crossline, offset and azimuth amplitude variations. A QC (Quality Control) tool known as "5D Leakage" is used to assess

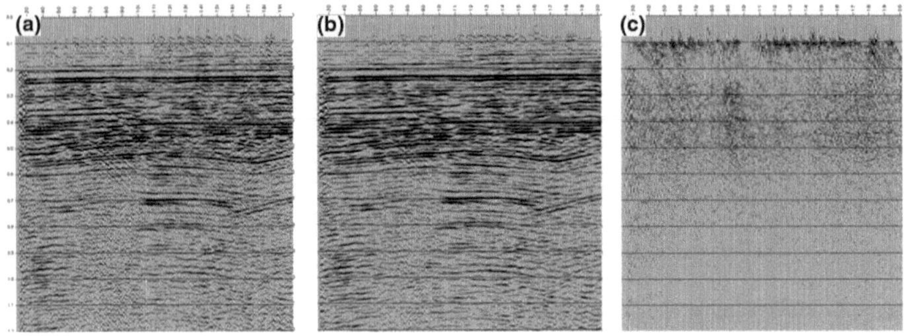

Fig. 4.3 5D interpolation and QC. **a** is the data stack without the 5D interpolation. **b** is data stack with the 5D interpolation in which the S/N ratio has been enhanced and the amplitude has been truly kept. **c** is an application of the QC tool of "5D Leakage" to inspect the 5D interpolation efficiency (Data same as in Fig. 4.2)

the accuracy of the 5D Interpolation by making a comparison between data with reconstructed geometry and true recorded data with the original geometry. Figure 4.3 shows the stack sections before and after 5D interpolation and QC "5D Leakage" (the data are the same as those shown in Fig. 4.2). Figure 4.3a shows the stacked data without the 5D interpolation, while Fig. 4.3b illustrates the stacked data with the 5D interpolation. Comparing Fig. 4.3a with Fig. 4.3b confirms that the S/N ratio of the data has been enhanced and the image is reliable and has true amplitudes. Figure 4.3c shows the QC of "5D Leakage" that verifies the 5D interpolation efficiency.

4.2.4 PS-Data Layer Stripping

4.2.4.1 Converted PS-Wave

In seismic exploration, the incident *P*-wave not only generates up-going and down-going PP-waves, but also it converts to an up-going PS reflection and down-going PS transmission once the *P*-wave impinges upon any reflector (Stewart et al. 2002, 2003). The PP and converted PS reflections would be fully present in multi-component data on the surface that usually contain one vertical component (PP-data) and two horizontal components (PS-data), i.e., they are 3C data. Figure 4.4 illustrates the paths of PP- and PS-waves and the mapping of the common middle point (CMP), the asymptotic conversion point (ACP), and the common conversion point (CCP). Sorting a CCP gather in PS-data is more complex than sorting a CMP gather in PP-data, because CCP traces also rely on the medium properties, whereas CMP traces only depend on the source and receiver geometry.

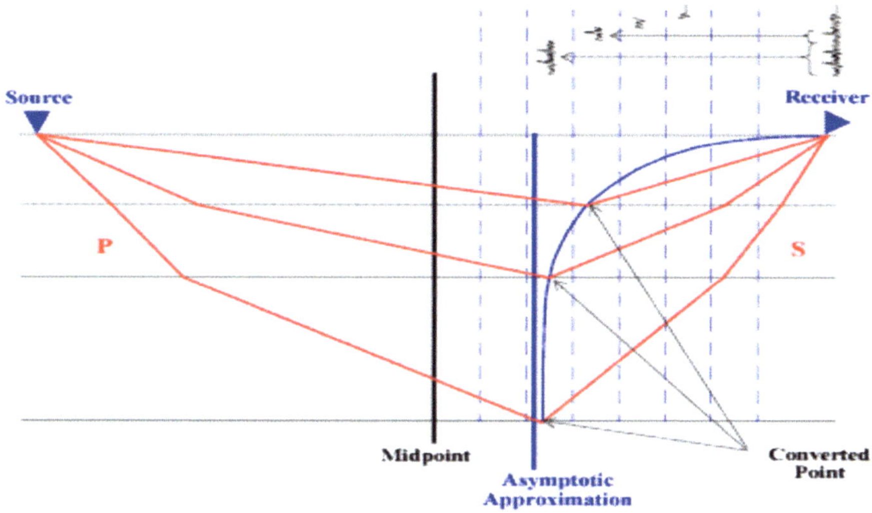

Fig. 4.4 Diagram for PP- and PS-wave paths and a mapping of the common middle point (CMP), asymptotic convert point (ACP), and common convert point (CCP)

4.2.4.2 Shear Wave Splitting

Due to the fact that a shear wave is always polarized orthogonally to the direction of wave propagation, a shear wave propagating in an azimuthally anisotropic medium with a number of unique properties can have a profound effect on the wavefront in terms of *S*-wave birefringence (Crampin 1985). Shear wave splitting techniques were first used for fracture detection by Alford (1986). At the fracture interface, a converted shear wave would split into a fast shear wave PS_1 that polarizes in and parallel to the direction of maximum stress and a slow shear wave PS_2 that is polarized in and parallel to the direction of minimum stress (Fig. 4.5). One may call this system (defined by the directions of maximum and minimum stress) the natural coordinate system. PS-wave splitting analysis is associated with the direction of the existing fracture. The split-converted shear waves PS_1 and PS_2 are orthogonally polarized with respect to each other and both are perpendicular to the direction of propagation and have different speeds that cause a time shift (Fang and Brown 1996).

In practice, the direction of the fracture in the natural system has an angle of (φ_0) with respect to the radial component *R*, or an angle of $(\varphi_0 + 90^\circ)$ with respect to the transverse component *T*. *R* and *T* are rotated from the two horizontal components in the acquisition system. The converted PS-wave that splits into a fast-wave PS_1 and a slow-wave PS_2 in the natural system satisfies the following equation:

$$\mathrm{PS}_1 = \mathrm{PS}(t)\cos\varphi_0, \tag{4.3a}$$

$$\mathrm{PS}_2 = \mathrm{PS}(t - \Delta t)\sin\varphi_0. \tag{4.3b}$$

As trigonometry in Fig. 4.5

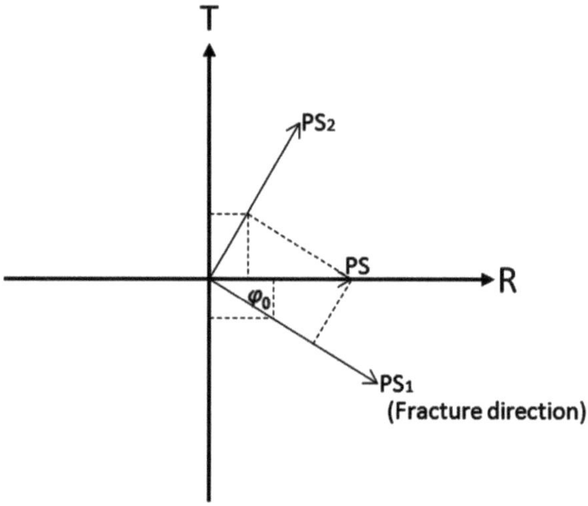

Fig. 4.5 A schematic of PS-wave splitting. The orientation of the fracture is φ_0 with respect to the radial component R. T denotes the transverse component. The split fast and slow PS-waves are $\boldsymbol{PS}_1$ and $\boldsymbol{PS}_2$, respectively. The direction of polarizations of $\boldsymbol{PS}_1$ and $\boldsymbol{PS}_2$ are orthogonal with respect to each other

$$\begin{bmatrix} \mathrm{PS}_1 \\ \mathrm{PS}_2 \end{bmatrix} = \begin{bmatrix} \cos\varphi_0 & -\sin\varphi_0 \\ \sin\varphi_0 & \cos\varphi_0 \end{bmatrix} \begin{bmatrix} R \\ T \end{bmatrix}, \tag{4.4a}$$

$$\mathcal{R}^{-1} \begin{bmatrix} \mathrm{PS}_1 \\ \mathrm{PS}_2 \end{bmatrix} = \begin{bmatrix} R \\ T \end{bmatrix}, \tag{4.4b}$$

where $\mathcal{R} = \begin{bmatrix} \cos\varphi_0 & -\sin\varphi_0 \\ \sin\varphi_0 & \cos\varphi_0 \end{bmatrix}$ is a rotator for the coordinate system rotation. And $\mathcal{R}^{-1} = \begin{bmatrix} \cos\varphi_0 & \sin\varphi_0 \\ -\sin\varphi_0 & \cos\varphi_0 \end{bmatrix}$.

Conversely, the fast and slow PS-waves can be obtained from the radial and transverse components by applying the rotator $\mathcal{R}$ if the fracture azimuth angle φ_0 is known:

$$\begin{bmatrix} \mathrm{PS}_1 \\ \mathrm{PS}_2 \end{bmatrix} = \mathcal{R} \begin{bmatrix} R \\ T \end{bmatrix} = \begin{bmatrix} R\cos\varphi_0 - T\sin\varphi_0 \\ R\sin\varphi_0 + T\cos\varphi_0 \end{bmatrix}. \tag{4.5}$$

4.2.4.3 Fracture Orientation

For PP-data, the orientation of existing fractures is often surmised from velocity anisotropy (Marrett et al. 2007). For PS-data, the methods of using the limitation of the amplitude value to determine the orientation of the fractures have been investigated (Garotta and Granger 1988; Bale et al. 2005). Given an arbitrary angle φ that is assumed to be the direction of the fracture with respect to the radial component in the R–T processing system, then

$$\mathrm{PS}_1(\varphi) = R\cos\varphi - T\sin\varphi, \tag{4.6a}$$

$$\mathrm{PS}_2(\varphi) = R\sin\varphi + T\cos\varphi. \tag{4.6b}$$

Substituting Eq. (4.4b) into Eqs. (4.6a) and (4.6b) the ratio of $\mathrm{PS}_2(\varphi)/\mathrm{PS}_1(\varphi)$ is

$$\frac{\mathrm{PS}_2(\varphi)}{\mathrm{PS}_1(\varphi)} = \frac{-\mathrm{PS}_1\cos\varphi_0\sin(\varphi_0-\varphi)+\mathrm{PS}_2\sin\varphi_0\cos(\varphi_0-\varphi)}{\mathrm{PS}_1\cos\varphi_0\cos(\varphi_0-\varphi)+\mathrm{PS}_2\sin\varphi_0\sin(\varphi_0-\varphi)}. \tag{4.7}$$

Substituting Eqs. (4.3a) and (4.3b) into Eq. (4.7) and if $\varphi = \varphi_0$, we have

$$\frac{\mathrm{PS}_2(\varphi_0)}{\mathrm{PS}_1(\varphi_0)} = \frac{\mathrm{PS}(t-\Delta t)\sin\varphi_0}{\mathrm{PS}(t)\cos\varphi_0}. \tag{4.8a}$$

Considering the energy of PS_1 and PS_2 only, then

$$\frac{|\mathrm{PS}_2(\varphi_0)|}{|\mathrm{PS}_1(\varphi_0)|} = \frac{|\mathrm{PS}(t-\Delta t)|}{|\mathrm{PS}(t)|}|\tan\varphi_0| = |\tan\varphi_0|, \tag{4.8b}$$

this φ_0 is the actual orientation of the fracture but it may be ambiguous with two distinct values at around ±90°. Usually, it is necessary to check the arrival time on the reflections by applying the two angle values to confirm that the fast shear wave has a smaller traveltime corresponding to the angle for the fracture direction.

4.2.4.4 Shear Wave Layer Stripping

If the time shift between fast and slow shear waves is big enough and not negligible, then the recorded waves on the surface are complicated by the destructive and constructive interference of fast and slow shear waves that degrade the quality of the data and render the data uninterpretable. Alford (1986) discovered an interesting operation, the layer stripping anisotropic correction, to overcome the problem by removing the time shift from shear wave spitting at a current anisotropic interface and the overburden anisotropic layers.

Figure 4.6 shows a schematic of a medium consisting of two subsurface layers. The deeper layer (Layer 1) is the target layer and the shallow layer (Layer 2) is a fractured layer (an azimuthally anisotropic layer). In Fig. 4.6, H_1 and H_2 represent two horizontal component data acquired on the surface in the acquisition system. PS_1 and PS_2 in layer 2 represent the fast and slow split waves from the converted PS-wave in the natural coordinate system, in which PS_1 is polarized in and parallel to the direction of the fracture (x_1) and PS_2 is polarized in and parallel to the direction perpendicular to the fracture (x_2). PS_1 and PS_2 are polarized orthogonally with respect to each other and propagate in the same direction with different speeds causing a time delay in the PS-data. In Fig. 4.6, shear wave splitting can be

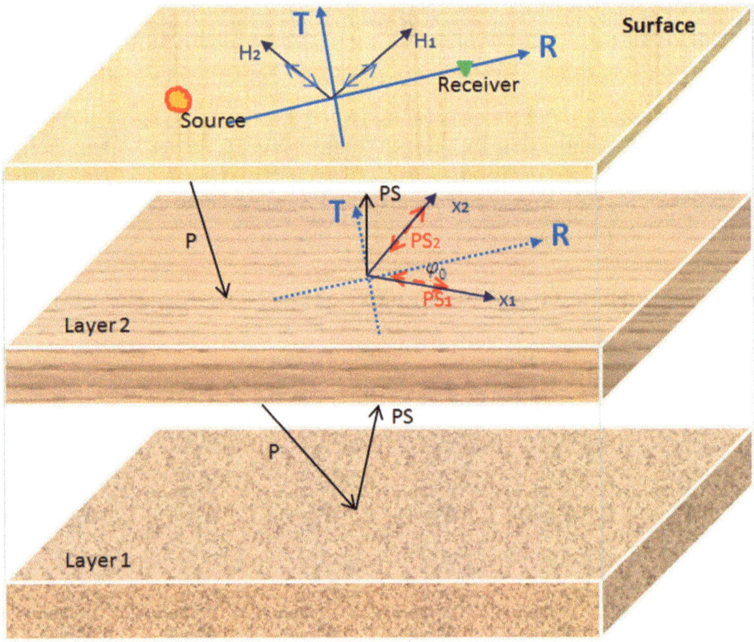

Fig. 4.6 Sketch of the shear wave layer stripping. $\mathbf{H}_1$ and $\mathbf{H}_2$ represent the two horizontal components of the converted data in the acquisition system. $\boldsymbol{PS}_1$ and $\boldsymbol{PS}_2$ at the anisotropic layer (Layer 2) symbolize the split fast and slow waves from the converted PS-wave in the natural system. The fast shear wave $\boldsymbol{PS}_1$ is polarized in and parallel to the direction of the fracture (x_1) and a slow shear wave $\boldsymbol{PS}_2$ is polarized in and parallel to the perpendicular direction of the fracture (x_2). $\boldsymbol{PS}_1$ and $\boldsymbol{PS}_2$ are orthogonally polarized respect to each other and propagate in the same direction with different speeds causing a time delay that appears in the PS-data

regarded as a forward process from layer 1 to the surface, while an inverse process is layer stripping from the surface to the target layer (layer 1).

In order to correctly image and interpret the target layer, the most important steps are as follows:

- Rotate the acquisition data in the acquisition system (H_1–H_2) into radial–transverse (R–T) data in the processing system.
- Rotate the radial–transverse (R–T) components of the pre-stack data into the (PS_1-PS_2) components of the natural (fracture) system (x_1–x_2) by applying Alford's rotation angle φ_0 in Eqs. (4.8a) and (4.8b).
- Determine a time lag for cross-correlation of the stacked traces of PS_1 and PS_2 in the analysis window.
- Apply the time shift to the pre-stack data to align all PS_2 traces to PS_1.
- Rotate the aligned PS_1 and PS_2 data back to the (R–T) components (normally named as ($R' - T'$) in the processing system.

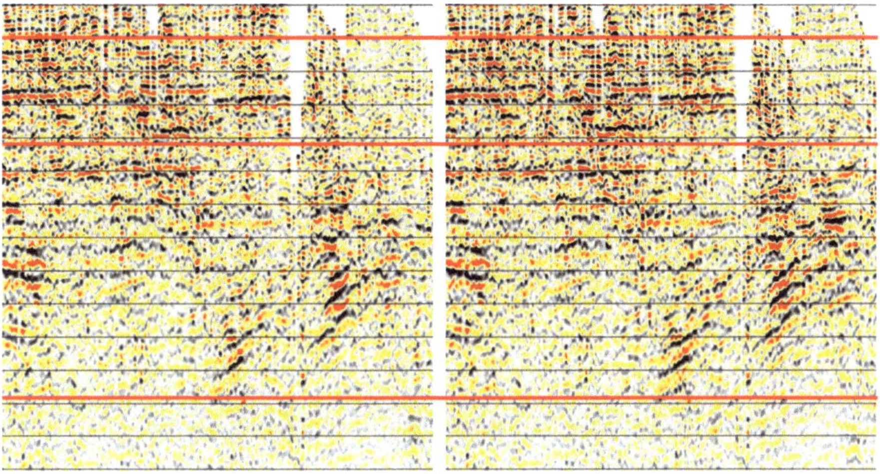

Fig. 4.7 Stack of shear wave layer stripping processing. The *left slice* is the radial component data stack without the layer stripping. The *right slice* is for the radial prime component data stack with the layer stripping. The layer stripping cascades two times at a shallow and deeper window. It is convincing that the layer stripping processing removed the time lag to enhance the S/N ratio and pronounce the reflection events

To significantly improve the reservoir image, these steps can be repeated layer by layer (for fractured anisotropic layers) with different analysis windows (Fig. 4.7). The explanation for this is that the different anisotropic (fractured) layers cause the time lag recorded in different time windows of the data. Meanwhile, an analysis of the amplitude and traveltime differences between the fast and slow converted shear waves may provide additional information to map subsequent fracture distributions in the reservoir. Figure 4.7 exhibits a result of the layer stripping that cascades two times of the layer stripping at shallow and deep windows. The left slice indicates the radial component data stacked without the layer stripping. The right slice is for the radial prime component data stacked with the layer stripping. It is convincing that the layer stripping processing removed the time lag to enhance the S/N ratio and improve the data quality of the R' component data.

4.3 AVO Equations for Fractured Media

When seismic waves propagate in the subsurface and encounter a fractured medium, the energy of the incident wave can be partially reflected and recorded on the surface seismic data (even if there is no impedance contrast around the fracture) that combine the response of the fracture and the host media (Chap. 3). In Chap. 3, the synthetic seismograms provide evidence that the fracture is a reflection generator and strongly affects the amplitudes of seismic traces, the traveltimes, and the $\frac{V_p}{V_s}$ ratio, and therefore the conventional AVO equations for inversion of the rock properties of the host media from seismic amplitudes will be inaccurate if the reservoir under investigation is fractured since the amplitudes inevitably are contaminated by fracture reflections. However, this issue is not recognized in the conventional AVO method.

AVO equations for a fractured medium should take the fracture issue into account and avoid its influences when the equations are used to invert for the elastic reflectivity of the host media from the input data that contain reflections from the fractured media.

4.3.1 Exact Reflectivity Equations for Horizontally Fractured Media

An incident *P*-wave ($P_1^{\grave{}}$), reflected *P*-wave ($P_1^{\grave{}}P_1'$), transmitted *P*-wave ($P_1^{\grave{}}P_2^{\grave{}}$), and converted *S*-waves ($P_1^{\grave{}}S_1'$ and $P_1^{\grave{}}S_2^{\grave{}}$), as well as the incidence angle (θ_1), transmission angle (θ_2), converted reflection angle (ϑ_1), converted transmission angle (ϑ_2), and a horizontal viscous fracture interface in the vertical *x*–*z* domain, are shown in Fig. 4.8. The black single arrows point in the directions of the wave

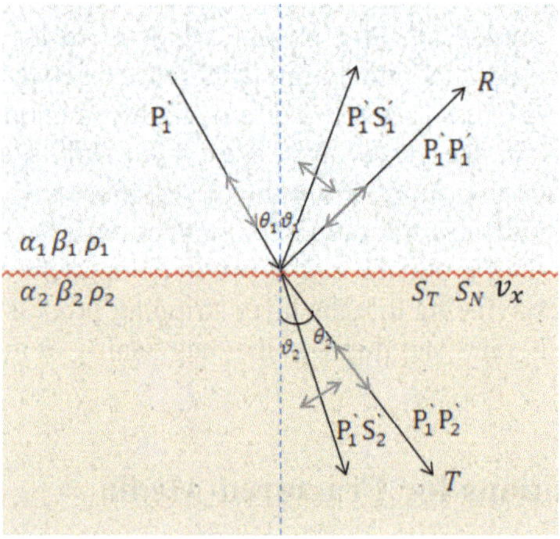

Fig. 4.8 Schematic of reflected and transmitted rays for an incident *P*-wave incident upon a fracture interface. The incident *P*-wave ($\mathbf{P}_1^{\grave{}}$), reflected *P*-wave ($\mathbf{P}_1^{\grave{}}\mathbf{P}_1'$), transmitted *P*-wave ($\mathbf{P}_1^{\grave{}}\mathbf{P}_2^{\grave{}}$), and converted *S*-waves ($\mathbf{P}_1^{\grave{}}\mathbf{S}_1'$ and $\mathbf{P}_1^{\grave{}}\mathbf{S}_2^{\grave{}}$), as well as incident angles ($\boldsymbol{\theta}_1$) transmission angle ($\boldsymbol{\theta}_2$), converted reflection angle ($\boldsymbol{\vartheta}_1$), and transmission angle ($\boldsymbol{\vartheta}_2$), are shown in the *x*–*z* domain. The *single black arrows* point in the direction of the wave propagation. The *double gray arrows* indicate the direction of the polarization of the waves

propagation. The gray double arrows indicate the directions of the polarizations of the waves. The rock properties of the upper and lower isotropic media are P-wave velocity ($\alpha_{1,2}$), S-wave velocity ($\beta_{1,2}$), and density ($\rho_{1,2}$). The normal compliance (S_{N}) and tangential compliance (S_{T}) of fracture and specific viscosity parameter (v_x) are shown in Fig. 4.8, too.

We use a harmonic incident plane P-wave, which can be expressed as $P_1^{\backprime} = Ae^{i\omega(\mathbf{s}\cdot\mathbf{x}-t)}\boldsymbol{d}$ (Krebes 2006, course notes), where A is the amplitude and is assumed to be unity, $\mathbf{s}\cdot\mathbf{x} = s_x x + s_y y + s_z z$, where $\mathbf{s}$ represents the slowness of the harmonic plane wave in the travel direction, and $\boldsymbol{d}$ stands for the wave polarization and its eigenvector component are $l_{\theta n}$, $m_{\theta n}$, $l_{\vartheta n}$, and $m_{\vartheta n}$, respectively, where $n = 1,2$ indicates the upper and lower medium (Ruger 2002). For a given frequency ω, the relationship between the PP-waves ($P_1^{\backprime}P_1'$, $P_1^{\backprime}P_2^{\backprime}$), the PS-waves ($P_1^{\backprime}S_1'$, $P_1^{\backprime}S_2^{\backprime}$), the angles ($\theta_{1,2}$, $\vartheta_{1,2}$), and the fracture parameters S_{T} and S_{N} at the horizontal viscous (v_x) interface satisfies the linear slip nonwelded contact boundary conditions (2.42) as (Appendix C)

$$\begin{bmatrix} l_{\theta1}\sin\theta_1 \\ m_{\theta1}\cos\theta_1 \\ (xx)_1\cos\theta_1 \\ (yy)_1\alpha_1 \end{bmatrix} = \begin{bmatrix} -l_{\theta1}\alpha_1 P & -m_{\vartheta1}\cos\vartheta_1 & l_{\theta2}\alpha_2 P - \frac{i\omega S_{\mathrm{T}}}{1-i\omega v_x S_{\mathrm{T}}}(xx)_2\cos\theta_2 & m_{\vartheta2}\cos\vartheta_2 - \frac{i\omega S_{\mathrm{T}}}{1-i\omega v_x S_{\mathrm{T}}}(rr)_2\beta_2 \\ m_{\theta1}\cos\theta_1 & -l_{\vartheta1}\beta_1 P & m_{\theta2}\cos\theta_2 - i\omega S_{\mathrm{N}}(yy)_2\alpha_2 & -l_{\vartheta2}\beta_2 P + i\omega S_{\mathrm{N}}(kk)_2\cos\vartheta_2 \\ (xx)_1\cos\theta_1 & (rr)_1\beta_1 & (xx)_2\cos\theta_2 & (rr)_2\beta_2 \\ -(yy)_1\alpha_1 & (kk)_1\cos\vartheta_1 & (yy)_2\alpha_2 & -(kk)_2\cos\vartheta_2 \end{bmatrix} \begin{bmatrix} P_1^{\backprime}P_1' \\ P_1^{\backprime}S_1' \\ P_1^{\backprime}P_2^{\backprime} \\ P_1^{\backprime}S_2^{\backprime} \end{bmatrix}, \tag{4.9}$$

where $i = \sqrt{-1}$,

$$l_{\theta n} = 1 + f\cos^2\theta_n(\delta + 2(\varepsilon - \delta)\sin^2\theta_n), \tag{4.10a}$$

$$m_{\theta n} = 1 - f\sin^2\theta_n(\delta + 2(\varepsilon - \delta)\sin^2\theta_n), \tag{4.10b}$$

$$l_{\vartheta n} = 1 + f\cos^2\vartheta_n(\delta + 2(\varepsilon - \delta)\sin^2\vartheta_n), \tag{4.10c}$$

$$m_{\vartheta n} = 1 - f\sin^2\vartheta_n(\delta + 2(\varepsilon - \delta)\sin^2\vartheta_n), \tag{4.10d}$$

$$(xx)_n = (l_{\theta n} + m_{\theta n})\rho_n\beta_n^2 P, \tag{4.10e}$$

$$(rr)_n = \rho_n(l_{\vartheta n} - (l_{\vartheta n} + m_{\vartheta n})\beta_n^2 P^2), \tag{4.10f}$$

$$(yy)_n = \rho_n\left[(l_{\theta n} - m_{\theta n})\alpha_n^2 P^2 + (m_{\theta n} - l_{\theta n}2\beta_n^2 P^2)\right], \tag{4.10g}$$

$$(kk)_n = \rho_n P\left((2\beta_n^2 - \alpha_n^2)m_{\vartheta n} + \alpha_n^2 l_{\vartheta n}\right), \tag{4.10h}$$

where $f = \frac{\alpha^2}{\alpha^2 - \beta^2}$, $n = 1, 2$. Equation (4.9) expresses reflection and transmission coefficients from the horizontally viscous fractured medium that formed a horizontal viscous fracture embedding in the anisotropic host medium. The formed medium is equivalent to TI medium that induced by the fracture and the host medium.

4.3.2 Approximate AVO Equations for Horizontally Fractured Media

The matrix on the left side of Eq. (4.9) is for the incident wave, and the matrix on the right is for the scattered wave and the reflection and transmission coefficients. Rewrite Eq. (4.9) as

$$M[P_1^{\backslash}P_1' \quad P_1^{\backslash}S_1' \quad P_1^{\backslash}P_2^{\backslash} \quad P_1^{\backslash}S_2^{\backslash}]^{\mathrm{T}} = N. \tag{4.11a}$$

According to Cramer's rule,

$$[P_1^{\backslash}P_1' \quad P_1^{\backslash}S_1' \quad P_1^{\backslash}P_2^{\backslash} \quad P_1^{\backslash}S_2^{\backslash}]^{\mathrm{T}} = \frac{\det(M_k)}{\det(M)}, \tag{4.11b}$$

where M_k is the matrix M and the k-th column of M has been replaced by the vector N. k = 1 corresponds to the first reflection/transmission coefficient, etc. Thus

$$P_1^{\backslash}P_1' = R_p(\theta) = \frac{\det(M_1)}{\det(M)} = \frac{\det\left(\begin{bmatrix} N_{11} & m_{12} & m_{13} & m_{14} \\ N_{21} & m_{22} & m_{23} & m_{24} \\ N_{31} & m_{32} & m_{33} & m_{34} \\ N_{41} & m_{42} & m_{43} & m_{44} \end{bmatrix}\right)}{\det\left(\begin{bmatrix} m_{11} & m_{12} & m_{13} & m_{14} \\ m_{21} & m_{22} & m_{23} & m_{24} \\ m_{31} & m_{32} & m_{33} & m_{34} \\ m_{41} & m_{42} & m_{43} & m_{44} \end{bmatrix}\right)}, \tag{4.12a}$$

$$P_1^{\backslash}S_1' = R_s(\theta) = \frac{\det(M_2)}{\det(M)} = \frac{\det\begin{pmatrix} m_{11} & N_{12} & m_{13} & m_{14} \\ m_{21} & N_{22} & m_{23} & m_{24} \\ m_{31} & N_{32} & m_{33} & m_{34} \\ m_{41} & N_{42} & m_{43} & m_{44} \end{pmatrix}}{\mathrm{ddet}\left(\begin{bmatrix} m_{11} & m_{12} & m_{13} & m_{14} \\ m_{21} & m_{22} & m_{23} & m_{24} \\ m_{31} & m_{32} & m_{33} & m_{34} \\ m_{41} & m_{42} & m_{43} & m_{44} \end{bmatrix}\right)}. \tag{4.12b}$$

Equations (4.12a) and (4.12b) are equations for the PP- and PS-wave reflection coefficients, respectively. It is straight forward to solve the linear algebraic Eq. (4.12a) with assumptions of the polarization vector $\boldsymbol{d} = 1$. Then the exact PP-wave reflectivity from the fractured VTI media can be formulated as (Appendix C)

$$R_p(\theta) = R_{\mathrm{w}}(\theta) + R_{\mathrm{non\_w}}(\theta), \tag{4.13}$$

where

$$R_{\mathrm{non\_w}}(\theta) = \left(\frac{i\omega S_{\mathrm{T}}}{1 - i\omega v_x S_{\mathrm{T}}}\right) R^{\mathrm{T}}_{\mathrm{non\_w}}(\theta) + (i\omega S_{\mathrm{N}}) R^{\mathrm{N}}_{\mathrm{non\_w}}(\theta). \tag{4.13a}$$

The PP reflection coefficient in Eq. (4.13) contains two items. The reflection coefficient R_{w} is caused by velocity or density contrasts in the host media that satisfy the assumption of perfectly welded contact. The reflection coefficient $R_{\mathrm{non\_w}}$ results from the fracture: the displacement discontinuity across the fracture generates the reflections. In Eq. (4.13a), the upper subscripts T and N indicate that the reflection coefficient is associated with the fracture tangential and normal compliance parameters S_{T} and S_{N}. For the conventional reflection coefficient R_{w}, it is assumed that all incidence and transmission angles are real and less than 90°, and it is the same approximation as given by Aki and Richards's AVO equation in Table 4.1:

$$R_{\mathrm{w}}(\theta) \approx \frac{1}{2\cos^2\theta} r_\alpha - 4\left(\frac{\beta}{\alpha}\right)^2 \sin^2\theta r_\beta + \frac{1}{2}\left(1 - 4\left(\frac{\beta}{\alpha}\right)^2 \sin^2\theta\right) r_\rho. \tag{4.14a}$$

For the fracture part $R_{\mathrm{non\_w}}$, it can be simplified into tangential and normal components with the same approximations that were used for R_{w}. So we have (Appendix C)

$$R^{\mathrm{T}}_{\mathrm{non_w}}(\theta) \approx 2\frac{1}{\rho}\mu^2 q_\alpha P^2 + \left(\frac{\mu}{q_\beta}P^2 - \frac{1}{4}\rho\beta \sec\vartheta \sec\theta^2\right) r_\alpha + 2\frac{\mu}{q_\beta}P^2 r_\beta + \left(\frac{\mu}{q_\beta}P^2 - \frac{1}{4}\rho\beta\sec\vartheta\right) r_\rho, \tag{4.14b}$$

$$R^{\mathrm{N}}_{\mathrm{non_w}}(\theta) \approx \frac{1}{2}\rho\alpha\sec\theta - \frac{2\mu}{q_\alpha}P^2 - \left(\mu^2\frac{q_\beta}{\rho}(1+\sin\theta^2)P^2\right) r_\alpha + \left(4\mu^2\frac{q_\beta}{\rho}P^2\right) r_\beta + \left(2\mu^2\frac{q_\beta}{\rho}P^2\right) r_\rho, \tag{4.14c}$$

where r_α, r_β, and r_β are compressional wave velocity, shear wave velocity, and density reflectivity, respectively; P is the ray parameter; and

$$r_\alpha = \frac{\Delta\alpha}{\alpha}, \quad r_\beta = \frac{\Delta\beta}{\beta}, \quad r_\rho = \frac{\Delta\rho}{\rho}, \tag{4.15a}$$

$$q_{\alpha_{1,2}} = \frac{\cos\theta_{1,2}}{\alpha_{1,2}}, \quad q_\alpha = \frac{q_{\alpha 1} + q_{\alpha 2}}{2}, \quad \Delta q_\alpha = q_{\alpha 2} - q_{\alpha 1}, \tag{4.15b}$$

$$q_{\beta_{1,2}} = \frac{\cos\vartheta_{1,2}}{\beta_{1,2}}, \quad q_\beta = \frac{q_{\beta 1} + q_{\beta 2}}{2}, \quad \Delta q_\beta = q_{\beta 2} - q_{\beta 1} \tag{4.15c}$$

$$\mu = \rho\beta^2. \tag{4.15d}$$

$R^{\mathrm{T}}_{\mathrm{non\_w}}(\theta)$ and $R^{\mathrm{N}}_{non\_\mathrm{w}}(\theta)$ have the same style as R_{w} in terms r_α, R_w and r_ρ. Thus Eq. (4.13) can be rewritten as (Appendix C)

$$\begin{aligned} R_p(\theta) &\approx R_{\mathrm{w}}(\theta) + \frac{i\omega S_{\mathrm{T}}}{1 - i\omega v_x S_{\mathrm{T}}} R^{\mathrm{T}}_{\mathrm{non\_w}}(\theta) + i\omega S_{\mathrm{N}} R^{\mathrm{N}}_{\mathrm{non\_w}}(\theta) \\ &\approx \frac{i\omega S_{\mathrm{T}}}{1 - i\omega v_x S_{\mathrm{T}}} T^{\mathrm{T}}(\theta) + i\omega S_{\mathrm{N}} N^{\mathrm{N}}(\theta) \\ &\quad + \left(A(\theta) + \frac{i\omega S_{\mathrm{T}}}{1 - i\omega v_x S_{\mathrm{T}}} A^{\mathrm{T}}_\alpha(\theta) + i\omega S_{\mathrm{N}} A^{\mathrm{N}}_\alpha(\theta)\right) r_\alpha \\ &\quad + \left(B(\theta) + \frac{i\omega S_{\mathrm{T}}}{1 - i\omega v_x S_{\mathrm{T}}} B^{\mathrm{T}}_\beta(\theta) + i\omega S_{\mathrm{N}} B^{\mathrm{N}}_\beta(\theta)\right) r_\beta \\ &\quad + \left(C(\theta) + \frac{i\omega S_{\mathrm{T}}}{1 - i\omega v_x S_{\mathrm{T}}} C^{\mathrm{T}}_\rho(\theta) + i\omega S_{\mathrm{N}} C^{\mathrm{N}}_\rho(\theta)\right) r_\rho. \end{aligned} \tag{4.16}$$

This is the new approximate AVO equation for the fractured medium with impedance contrast that divides the reflection coefficients into fracture reflection coefficients for the fracture and impedance contrast reflection coefficients for the host media. In other words, the all rock properties of the fractured medium with impedance contrast can be correctly inverted from the input data to delineate the reservoir characterizations. The host medium properties have a correct prediction because of the new AVO equation considered the input of seismic data possibly a combination of the reflection from the fracture issue. In Eq. (4.16)

$$T^{\mathrm{T}}(\theta) = 2\frac{1}{\rho}\mu^2 q_\alpha P^2, \quad N^{\mathrm{N}}(\theta) = \frac{1}{2}\rho\alpha\sec\theta - \frac{2\mu}{q_\alpha}P^2, \tag{4.17a}$$

$$A(\theta) = \frac{1}{2\cos^2\theta}, \quad A_\alpha^{\mathrm{T}}(\theta) = \frac{\mu}{q_\beta}P^2 - \frac{1}{4}\rho\beta\sec\vartheta\sec\theta^2,$$
$$A_\alpha^{\mathrm{N}}(\theta) = -(\mu^2\frac{q_\beta}{\rho}(1+\sin\theta^2)P^2), \tag{4.17b}$$

$$B(\theta) = -4\left(\frac{\beta}{\alpha}\right)^2\sin^2\theta, \quad B_\beta^{\mathrm{T}}(\theta) = 2\frac{\mu}{q_\beta}P^2, \quad B_\beta^{\mathrm{N}}(\theta) = 4\mu^2\frac{q_\beta}{\rho}P^2, \tag{4.17c}$$

$$C(\theta) = \frac{1}{2}\left(1 - 4\left(\frac{\beta}{\alpha}\right)^2\sin^2\theta\right), \quad C_\rho^{\mathrm{T}}(\theta) = \frac{\mu}{q_\beta}P^2 - \frac{1}{4}\rho\beta\sec\vartheta,$$
$$C_\rho^{\mathrm{N}}(\theta) = 2\mu^2\frac{q_\beta}{\rho}P^2. \tag{4.17d}$$

Figure 4.9 shows a comparison between the PP-wave exact solution and approximate solution of the reflection coefficients for the fractured medium with impedance contrast. The black line represents the exact solution from Eq. (4.9) as $v_x = 0°$ and $\boldsymbol{d} = 1$, while the red line is for the approximate solution from Eq. (4.16). The model parameters are the same as those in Fig. 4.9. This shows that the solution of the new approximate AVO Eq. (4.16) for fractured media closely matches the exact solution (Eq. 4.16) for incidence angles less than about 40°.

Let $r_\alpha = r_\beta = r_\rho = 0$ in Eq. (4.16). Then

$$R_p(\theta) \approx \left(\frac{i\omega S_{\mathrm{T}}}{1 - i\omega v_x S_{\mathrm{T}}}T^{\mathrm{T}}(\theta) + i\omega S_{\mathrm{N}}N^{\mathrm{N}}(\theta)\right). \tag{4.18}$$

Equation (4.18) clearly shows that the fracture parameters can be predicted from seismic data when the fractures are embedded in a uniform medium without impedance contrast. This can be used as an approach to verify the well-known

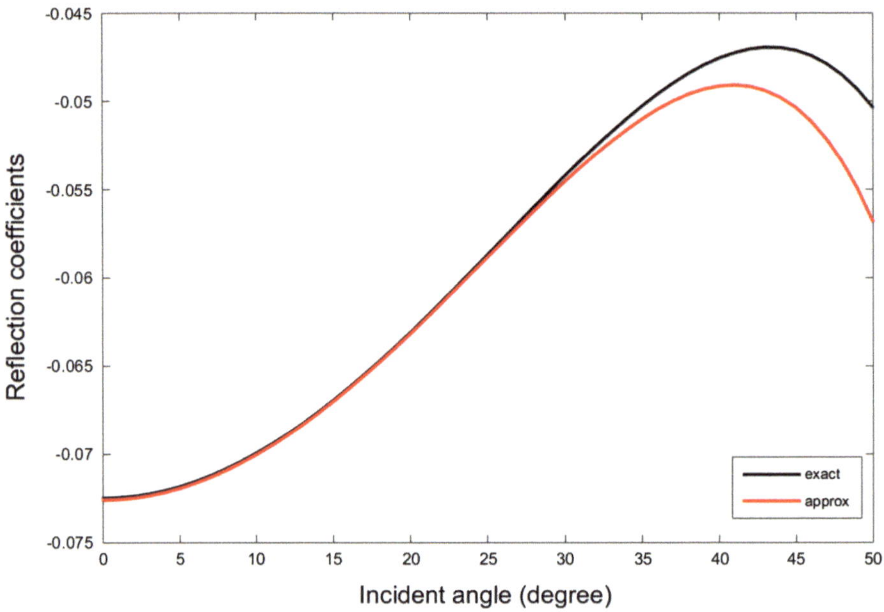

Fig. 4.9 PP reflection coefficients. The *black curve* is the exact solution for the fracture interface. The *red curve* is the approximation solution for the fracture interface. This illustrates that the approximate solution for the fracture interface is accurate in the conventional incidence angle range

parameters from the initial model. In particular, this equation provides a way to invert the rock fracture parameters, rather than a lithology variation from the seismic data to infer the reservoir characterization.

Let $S_{\mathrm{T}} = S_{\mathrm{N}} = 0$ for Eq. (4.16). Then

$$R_p(\theta) \approx A(\theta)r_\alpha + B(\theta)r_\beta + C(\theta)r_\rho. \tag{4.19}$$

Equation (4.19) states that the new AVO equation can be transformed into the conventional AVO equation to estimate the rock elastic reflectivity caused only by an impedance contrast without any fracture in the media.

4.4 AVO Inversion Algorithm for the Fractured Medium

Obtaining the geologic factors corresponding to the velocity reflectivity and density reflectivity is the ultimate objective of the AVO method. Equation (4.16) can be written in matrix format as

$$
\begin{bmatrix} R_p(\theta_1) \\ R_p(\theta_2) \\ \vdots \\ R_p(\theta_{m-1}) \\ R_p(\theta_m) \end{bmatrix} = \begin{bmatrix} i\omega\left(\frac{S_{\mathrm{T}}}{1-i\omega v_x S_{\mathrm{T}}} T^{\mathrm{T}}(\theta_1) + S_{\mathrm{N}} N^{\mathrm{N}}(\theta_1)\right) \\ i\omega\left(\frac{S_{\mathrm{T}}}{1-i\omega v_x S_{\mathrm{T}}} T^{\mathrm{T}}(\theta_2) + S_{\mathrm{N}} N^{\mathrm{N}}(\theta_2)\right) \\ \vdots \\ i\omega\left(\frac{S_{\mathrm{T}}}{1-i\omega v_x S_{\mathrm{T}}} T^{\mathrm{T}}(\theta_{m-1}) + S_{\mathrm{N}} N^{\mathrm{N}}(\theta_{m-1})\right) \\ i\omega\left(\frac{S_{\mathrm{T}}}{1-i\omega v_x S_{\mathrm{T}}} T^{\mathrm{T}}(\theta_m) + S_{\mathrm{N}} N^{\mathrm{N}}(\theta_m)\right) \end{bmatrix}
$$

$$
+ \begin{bmatrix} A(\theta_1) + i\omega\left(\frac{S_{\mathrm{T}}}{1-i\omega v_x S_{\mathrm{T}}} A_\alpha^{\mathrm{T}}(\theta_1) + S_{\mathrm{N}} A_\alpha^{\mathrm{N}}(\theta_1)\right) \\ A(\theta_2) + i\omega\left(\frac{S_{\mathrm{T}}}{1-i\omega v_x S_{\mathrm{T}}} A_\alpha^{\mathrm{T}}(\theta_2) + S_{\mathrm{N}} A_\alpha^{\mathrm{N}}(\theta_2)\right) \\ \vdots \\ A(\theta_{m-1}) + i\omega\left(\frac{S_{\mathrm{T}}}{1-i\omega v_x S_{\mathrm{T}}} A_\alpha^{\mathrm{T}}(\theta_{m-1}) + S_{\mathrm{N}} A_\alpha^{\mathrm{N}}(\theta_m - 1)\right) \\ A(\theta_m) + i\omega\left(\frac{S_{\mathrm{T}}}{1-i\omega v_x S_{\mathrm{T}}} A_\alpha^{\mathrm{T}}(\theta_m) + S_{\mathrm{N}} A_\alpha^{\mathrm{N}}(\theta_m)\right) \end{bmatrix}
$$

$$
\begin{matrix} B(\theta_1) + i\omega\left(\frac{S_{\mathrm{T}}}{1-i\omega v_x S_T} B_\beta^{\mathrm{T}}(\theta_1) + S_{\mathrm{N}} B_\beta^{\mathrm{N}}(\theta_1)\right) \\ B(\theta_2) + i\omega\left(\frac{S_{\mathrm{T}}}{1-i\omega v_x S_{\mathrm{T}}} B_\beta^{\mathrm{T}}(\theta_2) + S_{\mathrm{N}} B_\beta^{\mathrm{N}}(\theta_2)\right) \\ \vdots \\ B(\theta_{m-1}) + i\omega\left(\frac{S_{\mathrm{T}}}{1-i\omega v_x S_{\mathrm{T}}} B_\beta^{\mathrm{T}}(\theta_{m-1}) + S_{\mathrm{N}} B_\beta^{\mathrm{N}}(\theta_{m-1})\right) \\ B(\theta_m) + i\omega\left(\frac{S_{\mathrm{T}}}{1-i\omega v_x S_{\mathrm{T}}} B_\beta^{\mathrm{T}}(\theta_m) + S_{\mathrm{N}} B_\beta^{\mathrm{N}}(\theta_m)\right) \\ \cdots\cdots\cdots\cdots \\ C(\theta_1) + i\omega\left(\frac{S_{\mathrm{T}}}{1-i\omega v_x S_{\mathrm{T}}} C_\rho^{\mathrm{T}}(\theta_1) + S_{\mathrm{N}} C_\rho^{\mathrm{N}}(\theta_1)\right) \\ C(\theta_2) + i\omega\left(\frac{S_{\mathrm{T}}}{1-i\omega v_x S_{\mathrm{T}}} C_\rho^{\mathrm{T}}(\theta_2) + S_{\mathrm{N}} C_\rho^{\mathrm{N}}(\theta_2)\right) \\ \vdots \\ (\theta_{m-1}) + i\omega\left(\frac{S_{\mathrm{T}}}{1-i\omega v_x S_{\mathrm{T}}} C_\rho^{\mathrm{T}}(\theta_{m-1}) + S_{\mathrm{N}} C_\rho^{\mathrm{N}}(\theta_{m-1})\right) \\ C(\theta_m) + i\omega\left(\frac{S_{\mathrm{T}}}{1-i\omega v_x S_{\mathrm{T}}} C_\rho^{\mathrm{T}}(\theta_m) + S_{\mathrm{N}} C_\rho^{\mathrm{N}}(\theta_m)\right) \end{matrix} \Bigg] \begin{bmatrix} r_\alpha \\ r_\beta \\ r_\rho \end{bmatrix}, \tag{4.20}
$$

where m is the number of the offsets in a CDP gather. In Eq. (4.20), the matrix format declares that the new approximate AVO equation is applied to all the offsets of the CDP gather simultaneously. Rewrite Eq. (4.20) as a generalized linear inversion (GLI) problem, i.e.,

$$
Gx = d, \tag{4.21}
$$

where G is a linear operator depending on the geometry. x is the column vector for the unknown elastic reflectivity parameters r_α, r_β, and r_ρ. d is the column vector for the recorded seismic data. The least squares method is employed to solve the GLI Eq. (4.21). One obtains

$$x = [G^{T}G + \lambda I]^{-1}G^{T}d, \tag{4.21a}$$

the addition of a matrix λI tends to stabilize the calculation of $[G^{T}G]^{-1}$ (Lines and Treitel 1984). Equation (4.21a) can be used to solve all eight unknown elastic parameters for the fractured media.

4.5 Numerical Applications

4.5.1 Initial Model

A MATLAB code is implemented to do the numerical inversion and the results are based on the new AVO equations. The numerical model consists of the homogeneous isotropic host media with parameters of *P*-wave velocity $\alpha_{(1,2)} = 2850, 2800$ m/s; shear wave velocity $\beta_{(1,2)} = 1650, 1600$ m/s; density $\rho = 2.35$ g/cm^3; and a horizontal fracture with parameters of $S_{\mathrm{T}} = 0.127 \times 10^{-8}$ m/Pa and $S_{\mathrm{N}} = 0.269 \times 10^{-9}$ m/Pa (Fig. 4.10). This model focuses on

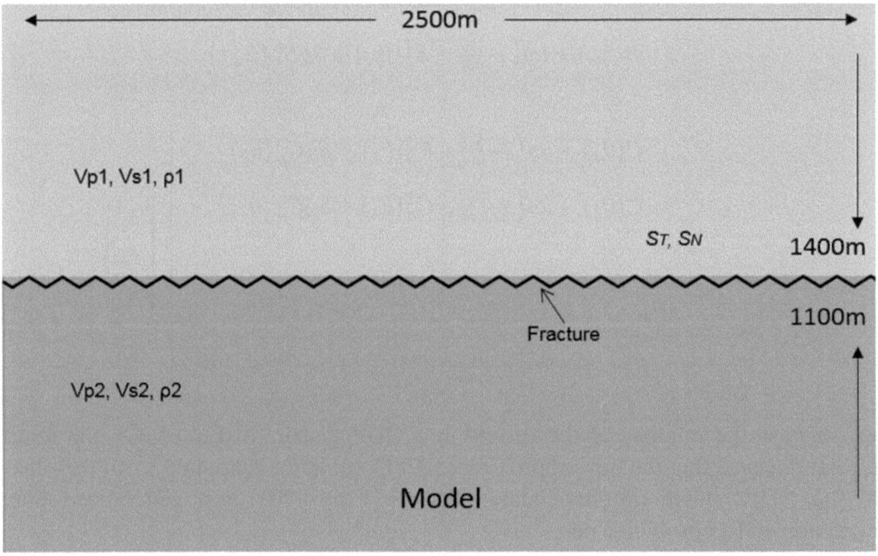

Fig. 4.10 AVO inversion model of the horizontally fractured media. A horizontal interface is embedded in homogeneous isotropic host media whose parameters are $\boldsymbol{\alpha}_{(1,2)} = 2800, 2850$ m/s; $\boldsymbol{\beta}_{(1,2)} = 1600, 1650$ m/s; the density is $\boldsymbol{\rho} = 2.35$ g/cm^3; and the fracture parameters are $\boldsymbol{S}_T = 0.127 \times 10^{-8}$ m/Pa and $\boldsymbol{S}_N = 0.269 \times 10^{-9}$ m/Pa

VTI-fractured medium with impedance contrast: a horizontal fracture embedded into the homogeneous host medium with impedance contrast. The input data AVO inversion for this fractured VTI medium have been synthetically generated by the forward modeling of the fracture described in Chap. 3.

4.5.2 Data Preconditioning

Deconvolution processing (using CREWES software) is applied to the input seismic data for AVO to remove the wavelet factors and flatten the amplitude spectrum. So the deconvolved input seismic data would have a uniform amplitude spectrum and higher resolution to more delicately represent the reflectivity of the fractured media. Figure 4.11 shows CDP gathers (with NMO applied) sorted from the shots with and without deconvolution (Fig. 4.11a, b) and their corresponding amplitude spectra (Fig. 4.11c). In Fig. 4.11c, the black curve represents the spectrum of the

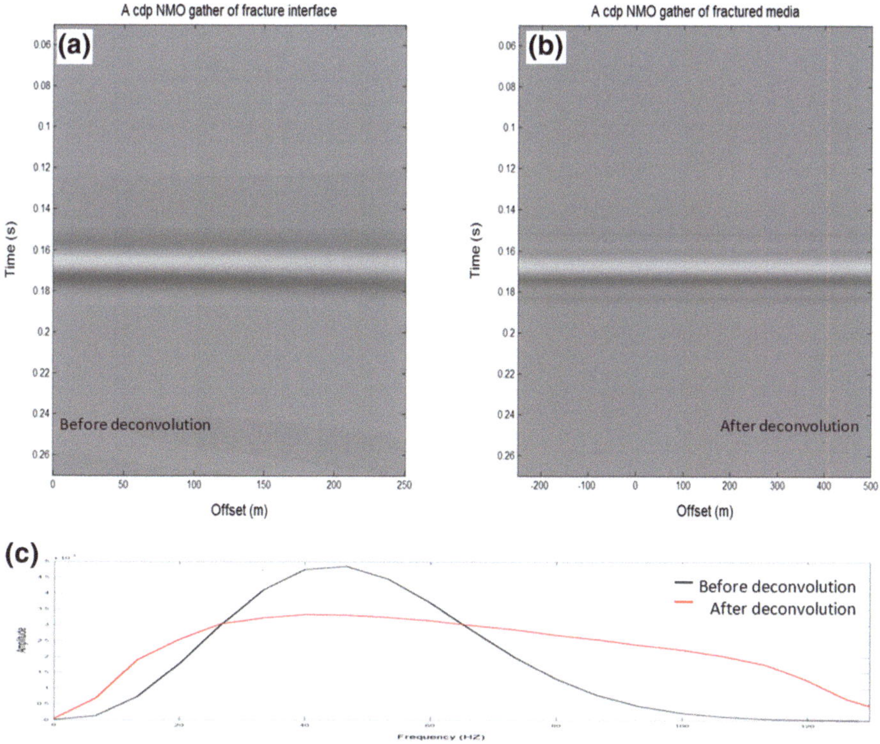

Fig. 4.11 Muted NMO gathers with (**b**) and without (**a**) deconvolution and their corresponding amplitude spectrum analyses for (**a**) and (**b**). In **c**, the *black curve* presents the data spectrum before deconvolution, and the *red curve* displays the data spectrum after deconvolution

data before deconvolution, and the red curve displays the spectrum of the data after deconvolution. On the other hand, the input data do not need noise attenuation and the 5D interpolation processes that are conducted in the processing flow for real seismic data.

4.5.3 Preparation of Input Data

As we discussed in Chap. 2, a fractured medium is a linear summation of a fracture and a host medium. Numerically, a similar relation applies to the amplitude of a CDP gather: Amplitude of the fractured medium with impedance contrast $\approx$ Amplitude of the fracture with medium 1 + Amplitude of the host medium with impedance contrast. This relationship is well numerically proved as shown in Fig. 4.12 which illustrates the reflection individually reflected from the fractured medium with impedance contrast (black), the fracture with uniform host medium 1 (red), and the host medium with impedance contrast (blue). The green line is a summation of the reflection of the fracture with uniform host medium 1 and the reflection of the host

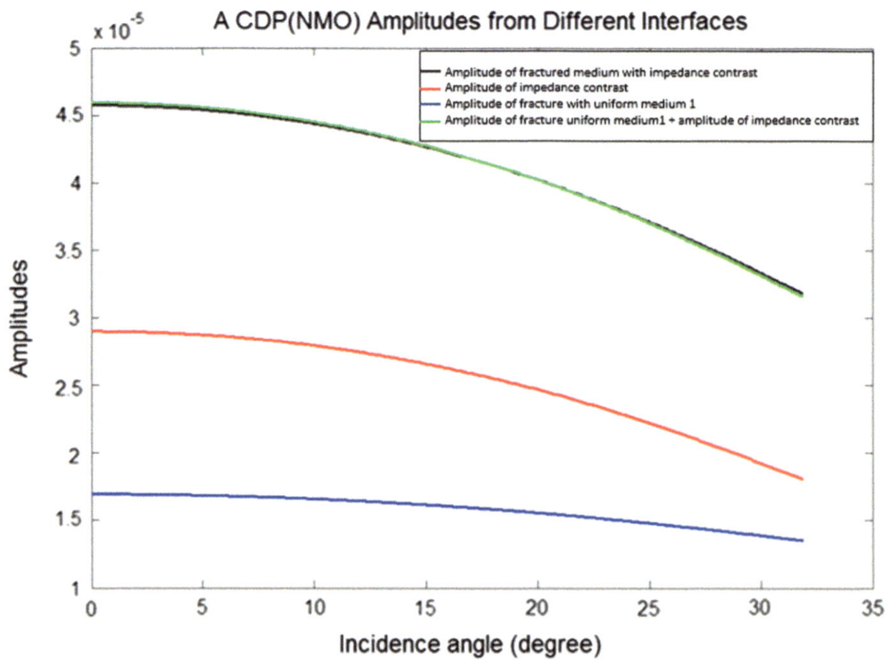

Fig. 4.12 CDP (with NMO applied) gathers reflectivity from different reflectors. The *black line* is the reflectivity *curve* for the fractured media with impedance contrast. The *red reflectivity curve* is reflected from a fracture. The *blue reflectivity curve* is generated from the homogeneous isotropic host media. The *green line* is made from the *red line* plus the *blue line*

medium with impedance contrast. It is not hard to observe that the summed reflection (green) approximately equals to the reflection of the fracture medium with impedance contrast (red + blue = green = black). It means that reflections of the pre-stack data from the fractured medium with impedance contrast can be divided into the reflection of the fracture and reflection of the host medium.

At this numerical application, the synthetic seismogram (shot records) is to be input seismic data.

Therefore, in order to truthfully invert the elastic reflectivity for the host medium, as a numerical study, the reflections caused by a fracture should be simulated by modeling a fracture embedding in the isotropic medium and then subtracted it from the input seismic data. For real data, the fracture parameters S_T and S_N are calculated from Thomson's anisotropic parameters using Eq. (3.39), while Thomson's anisotropic parameters can be evaluated from pre-stack seismic data because the fracture-induced anisotropy creates the fast and slow velocities leading to an uneven gather after NMO. Subsequently, the fracture reflection coefficients can be separated and inverted by Eq. (4.16) from the input seismic data.

Also, the parameters of the tangential and normal compliances of the fracture (S_T, S_N) can be predicted from seismic data by applying Eq. (4.18) if the host medium is a uniform isotropic medium.

4.5.4 Results Analysis

Figure 4.13 presents the AVO inversion results for the velocity reflectivity of the *P*- and *S*-wave for the host medium of the fractured medium with impedance contrast by applying the new approximate AVO Eq. (4.20) and implementing the GLI algorithm (Sect. 4.4). Furthermore, the input seismic data for AVO already

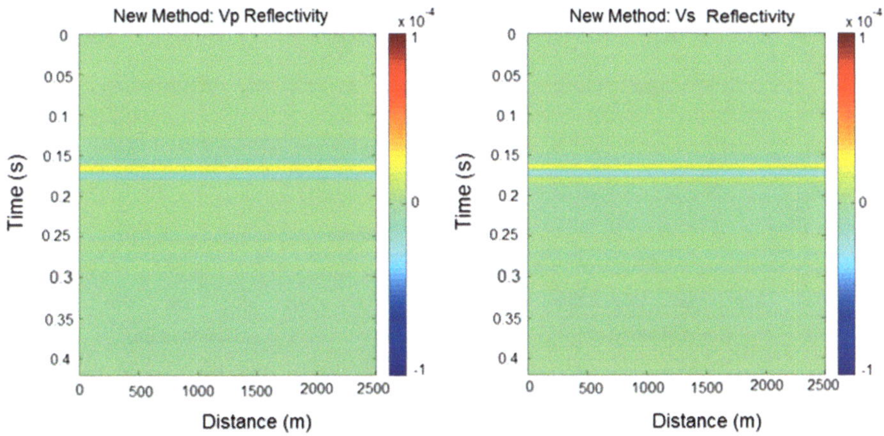

Fig. 4.13 Velocity reflectivity inversion. Velocity reflectivities of *P*- and *S*-waves of the host media are inverted from the fractured media seismic data by using the new AVO equations

have the fracture effects subtracted. At the time 0.1656 (s) in Fig. 4.13, the inverted interface correctly represents the interface of the host medium. Figure 4.14 shows the AVO inversion results for the velocity reflectivity of the *P*- and *S*-wave for the host medium of the fractured medium with impedance contrast too, but the results are obtained from the conventional AVO equations that do not consider fractures as reflectors. The conventional AVO equations invert the total reflectivity of the fractured medium with impedance contrast which is to be regarded as the velocity reflectivity for the host medium. With the same color bar, comparison of the two inversion results shown in Figs. 4.13 and 4.14 indicates that two AVO methods obtained different inverted velocity reflectivities for the same fractured medium with impedance contrast. Theoretically, the new approximate AVO equations in Fig. 4.13 provide more accurate inversion results than the conventional approximate AVO equations. Figure 4.15 shows the difference of the AVO results for the two methods. We can see that difference for the *S*-wave reflectivity is greater than the difference for the *P*-wave reflectivity. This observation agrees with the discussion in Chap. 3 that the *S*-wave is more sensitive to the fracture than the *P*-wave.

Figure 4.16 shows the AVO inversion results for the tangential and normal compliances of the fracture parameters from the fractured medium seismic data by applying the simplified new AVO Eq. (4.18). The input seismic data are generated only from the fracture reflector, and thus the host medium is a uniform homogeneous isotropic medium without impedance contrast. In other words, it is impossible to invert these fracture parameters by applying the conventional AVO equations because the conventional AVO equations are estimations for the reflectivity of the elastic properties.

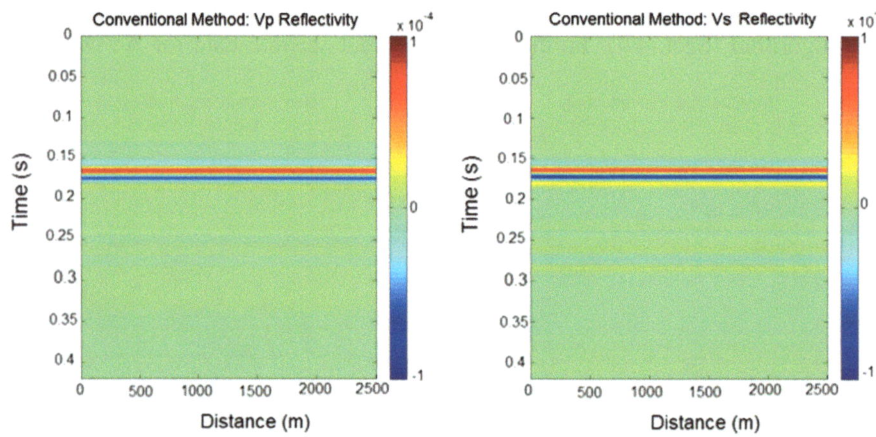

Fig. 4.14 Velocity reflectivity inversion. Velocity reflectivities of the host media are inverted from the fractured media seismic data using the conventional AVO equations

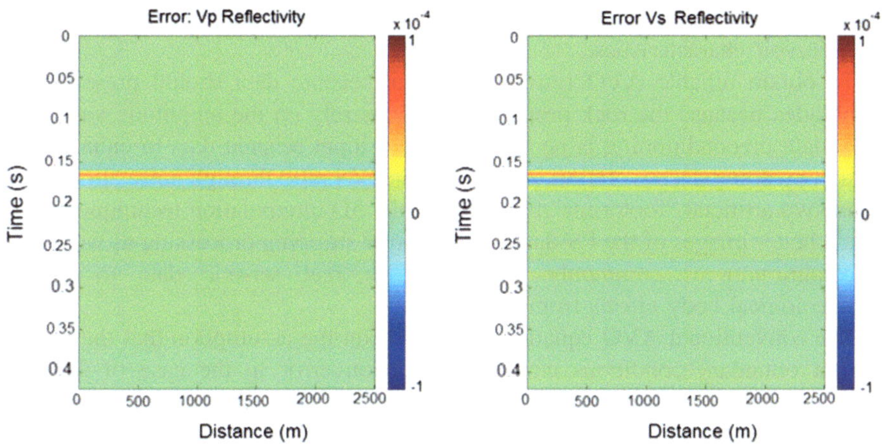

Fig. 4.15 Difference between *P*- and *S*-wave reflectivity inverted from new and conventional AVO equations

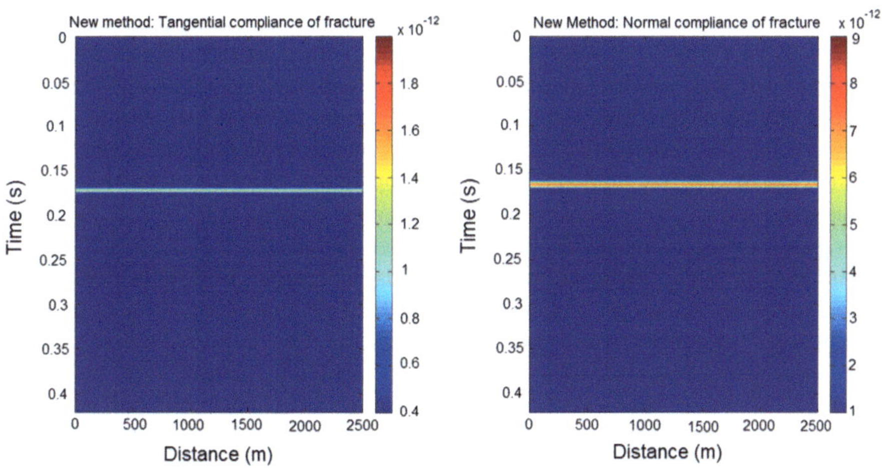

Fig. 4.16 Fracture parameters inversion. The tangential (*left*) and normal (*right*) compliances of the fracture parameters are inverted from the fractured seismic data by applying the simplified new AVO Eq. (4.18)

4.6 Conclusions

The equations of Zoeppritz (1919) describe the relationships between the incidence angles and the reflection and transmissions coefficients of PP- and PS-waves which is the basis of many approximate AVO inversion equations. In exploration, the AVO inversion method is widely used to estimate rock elastic reflectivity (*P*-waves,

S-waves, and density reflectivity) and then to infer more parameters that describe the reservoir characteristics.

To obtain reliable AVO results, the input seismic data should preserve true amplitudes because the rock property inversions rely on the amplitude variations. Thus data preconditioning is performed on the input seismic data to enhance resolution by doing deconvolution, to improve the S/N ratio through noise attenuation, to remove artificial "footprints" by applying the 5D interpolation technique, and to obtain better images of the PS-data through layer stripping processing as well. Data preconditioning is commonly performed for real seismic data to remove non-geological body effects from the data.

The conventional AVO equations are based on the assumption that the welded contact boundary conditions cause an inversion error in the case of fractured medium with impedance contrast. It is imperative that a method not only works for data from media with welded boundary conditions but is also adaptable to data from media with nonwelded boundary conditions. The equations expressing the relationships between the incidence angle of a plane wave and the reflection and transmission coefficients for the PP- and PS-waves have been derived. These equations are based on nonwelded boundary conditions. They have a pattern similar to that of the original Zoeppritz equations, but they take the fracture to be nonwelded contact interface. Also, these equations reduce to the original Zoeppritz equations for the welded media assumption, and also to equations only for reflection and transmission coefficients of the fracture in a uniform medium with the nonwelded contact assumption, as well as the equations reflection and transmission coefficients of the fractured medium with impedance contrast in the nonwelded media assumption. Thus the new equations can be adapted to describe subsurface geological bodies with two boundaries, as discussed above.

As we know, a fractured medium equals a fracture plus a host medium. Numerically, the reflection of fractured medium with impedance contrast approximately equals the reflection of the fracture plus the reflection of the host medium with impedance contrast. This has been numerically verified and shown in Fig. 4.12. Therefore, the reflection of the fracture can be separated and estimated from seismic data. Furthermore, subtracting the reflection of the fracture from the reflection of the fractured medium with impedance contrast gives the reflection of the host medium alone for correctly inverting for the rock properties of the host medium. New AVO equations have been developed for which the welded part works for the reflection from impedance contrast interfaces and the nonwelded part addresses the reflection from the fractures. The new AVO equations can reduce to other equations for simpler cases with certain assumptions. For example, the new AVO equations reduce to the conventional AVO equations when fracture parameters vanish, i.e., $S_{\mathrm{T}} = 0$, $S_{\mathrm{N}} = 0$, or to a fracture AVO equation when elastic reflectivity of the host medium is zero ($r_\alpha = r_\beta = r_\rho = 0$). Thus, an appropriate inversion of the elastic reflectivities of the host medium for the fractured medium with impedance contrast case should employ the new AVO equations because they provide a way to separate the fracture effects from the seismic data. In other words, the input data for inversion should not be contaminated by the fracture reflection when we attempt to invert for the elastic properties of the host medium. Similarly, the parameters of the tangential and normal compliances of the fracture can be determined from seismic data for a fractured medium by the fracture AVO Eq. (4.18).

Synthetic data for the fractured medium with impedance contrast have been used to invert for the rock properties of the host medium by applying both the new and conventional AVO equations respectively. The results show that the inversion results of the new AVO equations are theoretically and numerically more accurate than the results obtained from the conventional AVO equations (Figs. 4.13 and 4.14). The differences between the inversion results for these two AVO methods have been calculated and are shown in Fig. 4.15. In addition, the differences between the inversion results as shown in Fig. 4.15 indicate that the shear wave is more sensitive to the fracture than the *P*-wave. This observation agrees with the analysis of the seismograms in the forward modeling of Chap. 3. Therefore, one may conclude that the new AVO equations should be used to estimate the rock properties of the host medium in the fractured medium with impedance contrast case, so that we can correctly delineate a reservoir with fractures.

References

Aki, K., & Richards, P. G. (1980). *Quantitative seismology: Theory and methods* (Vol. 1). W.H. Freeman and Company.

Alford, R. M. (1986). *Shear data in the presence of azimuthal anisotropy*. Dilley, Texas. 56th SEG Annual Meeting, Expanded Abstracts, 86, S9.6, pp. 476–479.

Bale, R. A., Li, J., Mattocks, B., & Ronen, S. (2005). *Robust estimation of fracture directions from 3-D converted waves*. 75th SEG Annual Meeting, Expanded Abstracts, pp. 889–892.

Chaisri, S. (2002). *A study of a nonwelded contact interface: Exact and approximate formulas for P-SV reflection and transmission coefficients, and frequency domain raytracing* (Ph.D. thesis, The University of Calgary).

Chaisri, S., & Krebes, E. S. (2000). Exact and approximate formulas for P-SV reflection and transmission coefficients for a nonwelded contact interface. *Journal Geophysical Research, 105,* 28045–28054.

Choo, J., Downton, J., & Dewar, J. (2004). *A new and practical approach to noise and multiple attenuation*. 2004 CSEG National Convention.

Choo, J., Sudhakar, V. (2003). *LIFT: A new seismic processing technique to LIFT noise and multiples*. CSEG/CSPG Joint National Convention. Ecker, C., Dvorkin, J., & Nur, A. (1998). Sediments with gas hydrates: Internal structure from seismic AVO. *Geophysics*, 565–573.

Crampin, S. (1985). Evidence for aligned cracks in the Earth's crust. *First Break, 3*(3), 12–15.

Downton, J., Roure, B., & Hunt, L. (2011). Azimuthal Fourier coefficients. *CSEG Recorder, 36* (10), 22–36.

Fang, K., & Brown, R. J. (1996). *A new algorithm for the rotation of horizontal components of shear-wave seismic data* (Vol. 8). CREWES Research Report.

Fatti, J. L., Smith, G. C., Vail, P. J., Strauss, P. J., & Levitt, P. R. (1994). Detection of gas in sandstone reservoirs using AVO analysis: A 3-D seismic case history using the Geostack technique. *Geophysics, 59,* 1362–1376.

Gardner, G. H. F., Gardner, L. W., & Gregory, A. R. (1974). Formation velocity and density—The diagnostic basics for stratigraphic traps. *Geophysics, 39,* 770–780.

Garotta, R., & Granger, P. Y. (1988). *Acquisition and processing of 3C × 3-D data using converted waves*. 58th SEG Annual Meeting, Session S13.2.

Krebes, E. S. (2006). *Seismic theory and methods, course notes*. Calgary: The University of Calgary.

Lines, L. R., & Newrick, R. T. (2004). *Fundamentals of geophysical interpretation*. Tulsa, Oklahoma: Society of Exploration Geophysicists Publication.

Lines, L. R., & Treitel, S. (1984). Tutorial: A review of least-squares inversion and its application to geophysical problems. *Geophysical Prospecting, 32,* 159–186.

Margrave, G. F. (2006). *Methods of seismic data processing: Geophysics 557 course notes*. Calgary: University of Calgary.

Marrett, R., Laubach, S. E., & Olson, J. E. (2007). Anisotropy and beyond: Geologic perspectives on geophysical prospecting for natural fractures. *The Leading Edge, 26*(9), 1106–1111. Reprinted in *Fractured reservoirs: A compendium of influential papers*, 2008, AAPG.

Ruger, A. (2002). *Reflection coefficients and azimuthal AVO analysis in anisotropic media*. SEG geophysical monograph series number 10. Society of Exploration Geophysicists.

Shuey, R. T. (1985). A simplification of the Zoeppritz equations. *Geophysics, 50,* 609–614.

Smith, G. C., & Gidlow, P. M. (1987). Weighted stacking for rock property estimation and detection of gas. *Geophysical Prospecting, 35,* 993–1014.

Stewart, R. R., Gaiser, J. E., Brown, R. J., & Lawton, D. C. (2002). Converted-wave seismic exploration: Methods. *Geophysics, 67*(5), 1348–1363.

Stewart, R. R., Gaiser, J. E., Brown, R. J., & Lawton, D. C. (2003). Converted-wave seismic exploration: Applications. *Geophysics, 68*(1), 40–57.

Yilmaz, O. (1991). *Seismic data processing*. Society of Exploration Geophysicists (Fifth printing).

Zoeppritz, K. (1919). Erdbebenwellen VIIIB, On the reflection and propagation of seismic waves. *Göttinger Nachrichten, I,* 66.

Chapter 5
Conclusions and Future Work

Abstract This book is to develop a methodology for enhancing fracture detection and correctly delineating reservoirs with fractures. From the rock mechanics to seismic forward modeling then backward AVO inversion, the book deeply explores the mechanical formation of fractures and fractured media, and presents an enhanced fracture detection technique that uses a new finite-difference scheme to accurately model fractures and analyze the fracture response in seismic traveltime and amplitude, and then develops a method for accurate reservoir delineation by deriving new AVO fracture equations to correctly estimate the properties of the fractured medium, the host medium and fractured medium with impedance contrast.

Keywords Stress–strain tensors · Rock deformation · Stiffness compliances · Fracture formation · Fracture parameters · Backus average theory · Linear slip theory · Schoenberg–muir theory · Boundary conditions · Finite-difference (FD) modeling · Reflection/transmission coefficients · Inversion · Algorithm of the AVO inversion

5.1 Summary

In this book, we designed a methodology for the enhancement of fracture detection and correct delineation of a reservoir with fractures. The book mainly includes exploring fracture formation for understanding some key concepts related to the fractures, seismic forward modeling of a fracture for analyzing fracture responses in seismic data to enhance fracture detection, and the inversion of data for estimating all rock properties in order to correctly delineate a reservoir with fractures.

Through studying the key concepts of the related fracture formation, a fractured medium should undergo three deformation stages: elastic, ductile, and fracture. A fracture in the medium indicates that the rock strain exceeds the threshold when the rocks are continuously subjected to stress. In geoscience, a geological fracture relates to in situ stress and the fracture orientation is parallel to the direction of maximum compressive stress (Anderson 1951). Thus, most fractures in the

© Springer Nature Singapore Pte Ltd. 2018
X. Cui et al., *Seismic Forward Modeling of Fractures and Fractured Medium Inversion*, Springer Geophysics, DOI 10.1007/978-981-10-3584-5_5

reservoir are vertical or nearly vertical because the maximum compressive stress is a compression from the overburden deposits in the subsurface. Geological fractures are usually described by parameters of fracture length, width, density, opening, and orientation, while the fracture orientation is a more important parameter than the fracture opening because the seismic wavelength is greater than the fracture displacement (opening). Hence, seismic fracture modeling allows us to ignore the fracture shape and microstructure and uses a linear slip interface Schoenberg (1980) to simulate the fracture.

Seismic data with wider coverage and deeper penetration demonstrate a big potential for fracture detection, even though petrophysical well logs are regarded as the most reliable and are usually used to calibrate the seismic data, but the well logs are too sparse for fracture information. Therefore, in order to improve fracture detection and correctly delineate reservoirs with fractures, seismic fracture modeling and correct inversion of the properties of the fractured medium are addressed in this book.

The linear slip fracture model Schoenberg (1980) is based on the effective theory that mainly averages the multiple thin layers directly to make up the composite medium. Once the impedance contrast of a thin layer in the composite medium is much smaller than its host medium and its thickness is much smaller than a wavelength, this thin layer can be equivalent to a linear slip interface. Therefore, following the Schoenberg and Muir calculus (1989), a horizontally fractured medium is composed of a horizontal fracture embedded in an isotropic host medium, a vertically fractured medium is formed by a vertical fracture embedded in an isotropic host medium, and an orthogonally fractured medium is assembled from a horizontal fracture and a HTI host medium, or a vertical fracture and a VTI host medium, or two orthorhombic fractures embedded in an anisotropic host medium. The fact is that each composed fractured medium has five independent stiffnesses which indicate that the composed fractured medium possesses the medium properties of a transversely isotropic medium with a symmetric axis (TI). The boundary conditions constrain all waves at the boundary. For a fracture, they satisfy the nonwelded-contact boundary conditions in which the dynamic stresses of waves are continuous across the boundary, but the kinematic displacements are discontinuous across the boundary. This is different from welded-contact boundary conditions in which both stresses and displacements are continuous across the boundary.

Seismic forward modeling is a key connection between the model and the seismic response, and it can predict results, which enhance interpretation and inversion. It is clear that the homogenous finite-difference formulations of the elastic wave equation (Korn and Stöckl 1982; Slawinski 1999; Slawinski and Krebes 2002a, b) are sufficient for fracture seismic modeling because the boundary conditions can be imposed explicitly in this approach. We used the homogenous formulation approach by using additional fictitious nodes to derive new FD schemes for various fractured media, such as horizontally fractured medium, vertically fractured medium, and orthogonally fractured medium. In the FD grid for the new scheme, only one grid boundary in a cell is equivalent to a fracture interface that satisfies the nonwelded-contact boundary conditions, and the other three grid boundaries satisfy the welded-contact boundary conditions, whereas Slawinski and Krebes (2002a, b) allowed all four grid boundaries to be in nonwelded contact, for simplicity of coding and for flexibility in modeling. The new finite-difference scheme states that the normal equation of wave motion governs the wave propagation in the host medium, and that the nonwelded boundary conditions constrain the waves at the fracture interface when the waves propagate in the fractured medium.

The new finite-difference scheme was used with a Ricker wavelet source to generate the synthetic seismograms for the different fractured medium by implementing a MATLAB script. The seismograms indicate that the fractures are detectable and visible because the fractures strongly affect the seismic wave propagation as they give rise to reflections, even the medium does not have impedance contrasts. Also, analysis of the PP and PS amplitudes in the seismograms can identify the direction of the fractures: the seismograms of the horizontally fractured medium show that PP-wave amplitudes dominate in the z-component data, while PS-wave amplitudes are stronger than PP-wave amplitudes in the x-component of the seismograms. Conversely, in a seismogram for the vertically fractured medium, PP-wave amplitudes are stronger than PS-wave amplitudes in the x-component, while PS-wave amplitudes dominate in the z-component of the seismogram. Thus, the amplitudes of PP- and PS-waves in the x- and z-components of the seismograms can be used to detect the direction of the fracture. In addition, the seismograms illustrate that the fractures induce anisotropy. For example, in the horizontally fractured medium (VTI), the traveltime of a wave propagating parallel to the horizontal fracture is shorter than the traveltime of a wave propagating normally to it. In the vertical fractured medium, the wave propagation in the near fracture plane has a shorter traveltime than the wave propagation in the plane normal to the fracture plane. In other words, the vertically fractured medium demonstrates the following feature of HTI anisotropy: the medium properties vary with azimuth (AVAZ), i.e., the wave propagating in a different plane with a different azimuth has a different traveltime (velocity). Therefore, seismic modeling of fractures and analysis of the fracture response in the seismic signatures can highlight the existing fractures and enhance fracture detection.

Using any available results to infer the lithology and fluid properties in the reservoir is the ultimate goal for oil and gas exploration (Lines and Newrick 2004). AVO (amplitude variation with offset) inversion attempts to use the amplitudes of the available surface seismic data to estimate the reflectivity of the density, P-wave velocity and S-wave velocity of the earth model. Since the 1960s, geophysicists have discovered that gas deposits are related to amplitude anomalies on stacked sections, known as “bright spots”, and many geoscientists have been aware that surface-recorded seismic amplitudes can be related to rock properties by approximations to the solutions to the Zoeppritz equations (1919). However, those AVO equations are completely based on an assumption of a perfectly welded-contact medium regardless of whether or not fractures exist in the medium.

AVO inversion is based on the amplitude variation over a range of the offsets from the surface seismic data. Therefore, the seismic data must accurately preserve the true amplitudes corresponding to true geological factors, rather than contain useless signs of non-geological bodies and artifacts from acquisition. Thus we conducted some preconditioning processing, such as deconvolution, noise attenuation, 5D interpolation, and shear-wave layer stripping and showed their effects in real seismic data.

New reflection and transmission equations with parameters of tangential and normal fracture compliances (S_T and S_N) and viscosity parameter (ν_x) are presented based on nonwelded-contact boundary conditions. We numerically computed and presented a graphic which convincingly showed that the reflection from a fractured medium with impedance contrast can be approximately decomposed into a fracture-caused reflection and a reflection caused by the host medium, and vice versa. The equations have a pattern similar to that of the original Zoeppritz equations, but they take the fracture factor into account. The equations reduce to the original Zoeppritz equations with the assumption of welded contact, i.e., when the tangential and normal compliance vanish, and they reduce to the fractured AVO equations when the medium has no impedance contrast. Thus, the new equations can describe subsurface geological bodies with both welded and nonwelded-contact boundary conditions.

We derived some new approximate AVO equations for inversion of the horizontally fractured medium. The new AVO equations consist of the reflection coefficients of the welded-contact part caused by an impedance contrast at the interface and the reflection coefficients of the nonwelded-contact part caused by the fracture. Thus, an accurate inversion of the elastic reflectivities of the host medium of the fractured medium with impedance contrast should apply the new AVO equations because they provide a way to separate the fracture effects from the seismic data. In other words, the input data for AVO inversion should avoid contamination of the fracture reflection when we attempt to invert for the elastic properties of the host medium. Also, the fracture properties of the fractured medium for which the fracture is embedded in a uniform isotropic host medium can be estimated by using the new AVO inversion equations, whereas with conventional AVO inversion it is hard to achieve this because the conventional AVO inversion is used to invert the medium reflectivity with the assumption of a welded-contact medium. Therefore, the new AVO inversion equations can be used to invert for the properties of either a welded-contact or nonwelded-contact medium, and especially for media related to fractures.

Finally, we have shown the results of inversion for the properties of the host medium for the fractured medium with impedance contrast by applying the new AVO equations, and have also displayed the incorrect inversion results by using the conventional AVO equations to invert the properties of the host medium. Similarly, we illustrated an incorrect result for the fracture properties by applying the conventional AVO equations, because the conventional AVO equations only have the reflectivity corresponding to the host medium, and neglect the fracture effect. Also, we demonstrated the fracture properties that are estimated from the reduced form of the new AVO equations. Therefore, applying the correct AVO inversion equations allows one to estimate all rock properties and one is able to delineate the reservoir truthfully.

5.2 Future Work

Throughout this book, it was assumed that the fractures are unfilled. However, the infilled material strongly affects the boundary conditions and the fracture response in the seismic data. In future, we will focus on the research of the field cases where the fluids and some of highly viscous hydrocarbons (e.g., bitumen) are in the infill fractures.

Uncertainty is ubiquitous in reservoir characterization, therefore, we will consider an uncertain analysis with a different fracture structure to model and study their response in the seismic data.

The numerical forward modeling and inversion have been presented in this book. We will study the different models, e.g., an isotopic medium over the sets of the vertical fracture, and apply real seismic data to analyze the fracture representations, and invert all of the rock properties related to the fractured medium by using the new AVO equations.

5.3 Conclusions

In this book, a methodology for understanding the fracture mechanism, enhancing the fracture detection and for correctly delineating the reservoirs with the fractures has been developed. It can assist reservoir and mining engineers and geoscientists to optimize hydrocarbon reservoir, well drilling performance, and safety mining.

References

Anderson, E. M. (1951). The dynamics of faulting and dyke formation with application to britain. Oliver & Boyd: Edinburgh.

Korn, M., & Stöckl, H. (1982). Reflection and transmission of love channel wave at coal seam discontinuities computed with a finite difference method. *Journal of Geophysics, 50,* 171–176.

Lines, L. R., & Newrick, R. T. (2004). *Fundamentals of geophysical interpretation: society of exploration geophysicists publication*. Oklahoma: Tulsa.

Schoenberg, M. (1980). Elastic wave behavior across linear slip interfaces. *Journal of the Acoustical Society of America, 68,* 1516–1521.

Schoenberg, M., & Muir, F. (1989). A calculus for finely layered anisotropic media. *Geophysics, 54,* 581–589.

Slawinski, R. A. (1999). Finite-difference modeling of seismic wave propagation in fractured media: Ph.D. thesis, The University of Calgary.

Slawinski, R. A., & Krebes, E. S. (2002a). Finite-difference modeling of SH-wave propagation in nonwelded contact media. *Geophysics, 67,* 1656–1663.

Slawinski, R. A., & Krebes, E. S. (2002b). The homogeneous finite difference formulation of the P-SV wave equation of motion. *Studia Geophysica et Geodaetica* (Impact Factor: 0.75). 01/2002; *46*(4), 731–751. DOI:10.1023/A:1021133606779

Zoeppritz, K. (1919). Erdbebenwellen VIIIB, On the reflection and propagation of seismic waves: Göttinger Nachrichten, I, 66.

Appendix A
Moduli Calculation for Fractured Media

A.1 Schoenberg and Muir (1989) Calculus Theory

The matrix for the rock physical parameters (the stiffness matrix) can be divided into four sub-matrices, and the relationship between stress and strain for the equivalent layered (from to bottom) medium is

$$\begin{bmatrix} \overline{\sigma}_{\mathrm{T}} \\ \sigma_{\mathrm{N}} \end{bmatrix} = \begin{bmatrix} C_{\mathrm{TT}} & C_{\mathrm{TN}} \\ C_{\mathrm{TN}}^{\mathrm{T}} & C_{\mathrm{NN}} \end{bmatrix} \begin{bmatrix} \varepsilon_{\mathrm{T}} \\ \overline{\varepsilon}_{\mathrm{N}} \end{bmatrix}, \tag{A.1}$$

where $\overline{\sigma}_{\mathrm{T}}$, σ_{N}, ε_{T} , and $\overline{\varepsilon}_{\mathrm{N}}$ are defined in Eqs. (2.22) and (2.23) in Chap. 2, and

$$\overline{\sigma}_{\mathrm{T}} = \left[\langle C_{\mathrm{TT}}\rangle - \left\langle C_{\mathrm{TN}} C_{\mathrm{NN}}^{-1} C_{\mathrm{NT}}\right\rangle\right]\varepsilon_{\mathrm{T}} + \left\langle C_{\mathrm{TN}} C_{\mathrm{NN}}^{-1}\right\rangle \sigma_{\mathrm{N}}, \tag{A.2a}$$

$$\overline{\varepsilon}_{\mathrm{N}} = -\left\langle C_{\mathrm{NN}}^{-1} C_{\mathrm{NT}}\right\rangle \varepsilon_{\mathrm{T}} + \left\langle C_{\mathrm{NN}}^{-1}\right\rangle \sigma_{\mathrm{N}}, \tag{A.2b}$$

$$C_{\mathrm{NN}} = \left\langle C_{\mathrm{NN}}^{-1}\right\rangle^{-1}, \tag{A.2c}$$

$$C_{\mathrm{TN}} = \left\langle C_{\mathrm{TN}} C_{\mathrm{NN}}^{-1}\right\rangle C_{\mathrm{NN}}, \tag{A.2d}$$

$$C_{\mathrm{TT}} = \langle C_{\mathrm{TT}}\rangle - C\left\langle {}_{\mathrm{TN}} C_{\mathrm{NN}}^{-1} C_{\mathrm{NT}}\right\rangle + \left\langle C_{\mathrm{TN}} C_{\mathrm{NN}}^{-1}\right\rangle C_{\mathrm{NN}} \left\langle C_{\mathrm{NN}}^{-1} C_{\mathrm{NT}}\right\rangle, \tag{A.2e}$$

The bracket $\langle\cdot\rangle$ meansa thickness-weighted average. σ_{N}, ε_{T} are layering independent. The elements of the stiffness can be mapped into an Abelian group G with 5 elements that includes two scalars and three 3×3 matrices.

$$G = [\,g(1),\quad g(2),\quad g(3),\quad g(4),\quad g(5)\,]^{\mathrm{T}}. \tag{A.3}$$

Each element of the group relates to the physical parameters as

© Springer Nature Singapore Pte Ltd. 2018

X. Cui et al., *Seismic Forward Modeling of Fractures and Fractured Medium Inversion*, Springer Geophysics, DOI 10.1007/978-981-10-3584-5

$$\begin{bmatrix} H \\ \rho \\ C_{\mathrm{NN}} \\ C_{\mathrm{TN}} \\ C_{\mathrm{TT}} \end{bmatrix} \rightarrow \begin{bmatrix} H \\ H\rho \\ HC_{\mathrm{NN}}^{-1} \\ HC_{\mathrm{TN}}C_{\mathrm{NN}}^{-1} \\ H\left[C_{\mathrm{TT}} - C_{\mathrm{TN}}C_{\mathrm{TT}}^{-1}C_{\mathrm{NT}}\right] \end{bmatrix} = \begin{bmatrix} g(1) \\ g(2) \\ g(3) \\ g(4) \\ g(5) \end{bmatrix}, \tag{A.4}$$

H is total thickness of the medium. ρ is average density of the medium. Equation (A.4) can be used for the Abelian group of the host medium.

$$\begin{bmatrix} H \\ H\rho \\ HC_{\mathrm{NN}}^{-1} \\ HC_{\mathrm{TN}}C_{\mathrm{NN}}^{-1} \\ H\left[C_{\mathrm{TT}} - C_{\mathrm{TN}}C_{\mathrm{TT}}^{-1}C_{\mathrm{NT}}\right] \end{bmatrix} = \begin{bmatrix} g_H(1) \\ g_H(2) \\ g_H(3) \\ g_H(4) \\ g_H(5) \end{bmatrix}. \tag{A.5a}$$

The host medium is fractured, the physical parameters of the deformed fracture interface are mapped approximately as

$$\begin{bmatrix} 0 \\ 0 \\ HC_{\mathrm{NN}}f^{-1} \\ 0 \\ 0 \end{bmatrix} = \begin{bmatrix} g_f(1) \\ g_f(2) \\ g_f(3) \\ g_f(4) \\ g_f(5) \end{bmatrix}, \tag{A.5b}$$

Equation (A.5b) shows that all elements of the fracture interface vanish except for $g_f(3)$ which is known as the fracture compliance matrix (Eq. 2.23). The Abelian group of a fractured medium is equivalent to the Abelian group of the host medium plus that of the fracture.

$$\begin{bmatrix} g(1) \\ g(2) \\ g(3) \\ g(4) \\ g(5) \end{bmatrix} = \begin{bmatrix} g_H(1) \\ g_H(2) \\ g_H(3) \\ g_H(4) \\ g_H(5) \end{bmatrix} + \begin{bmatrix} g_f(1) \\ g_f(2) \\ g_f(3) \\ g_f(4) \\ g_f(5) \end{bmatrix} = \begin{bmatrix} g_H(1) \\ g_H(2) \\ g_H(3) + g_f(3) \\ g_H(4) \\ g_H(5) \end{bmatrix}. \tag{A.6}$$

One may invert the calculated elements of the group in Eq. (A.6) to obtain the rock physical parameters, as follows:

$$\begin{bmatrix} g(1) \\ g(2)/g(1) \\ g(1)g(3)^{-1} \\ g(4)g(3)^{-1} \\ \left[g(5) + g(4)g(3)^{-1}g(4)^{\mathrm{T}}\right]/g(1) \end{bmatrix} = \begin{bmatrix} H_H \\ \rho \\ C_{\mathrm{NN}} \\ C_{\mathrm{TN}} \\ C_{\mathrm{TT}} \end{bmatrix}. \tag{A.7}$$

A.2 Moduli Calculation of the Horizontally Fractured Medium

The horizontally fractured medium consists of a horizontal fracture interface and an isotropic homogeneous host medium. The stiffness of the isotropic homogeneous host medium can be specified as

$$C_{\mathrm{TT}\_H} = \begin{bmatrix} C_{11H} & C_{12H} & 0 \\ C_{21H} & C_{22H} & 0 \\ 0 & 0 & C_{66H} \end{bmatrix} = \begin{bmatrix} \lambda+2\mu & \lambda & 0 \\ \lambda & \lambda+2\mu & 0 \\ 0 & 0 & \mu \end{bmatrix}, \tag{A.8a}$$

$$C_{\mathrm{NN}\_H} = \begin{bmatrix} C_{33H} & 0 & 0 \\ 0 & C_{44H} & 0 \\ 0 & 0 & C_{55H} \end{bmatrix} = \begin{bmatrix} \lambda+2\mu & 0 & 0 \\ 0 & \mu & 0 \\ 0 & 0 & \mu \end{bmatrix}, \tag{A.8b}$$

$$C_{\mathrm{TN}\_H} = \begin{bmatrix} C_{13H} & 0 & 0 \\ C_{23H} & 0 & 0 \\ 0 & 0 & 0 \end{bmatrix} = \begin{bmatrix} \lambda & 0 & 0 \\ \lambda & 0 & 0 \\ 0 & 0 & 0 \end{bmatrix}. \tag{A.8c}$$

The elements of the group for the fractured medium are

$$g(3) = H \begin{bmatrix} \frac{1}{C_{33H}} + S_{\mathrm{N}} & 0 & 0 \\ 0 & \frac{1}{C_{44H}} + S_{\mathrm{T}} & 0 \\ 0 & 0 & \frac{1}{C_{55H}} + S_{\mathrm{T}} \end{bmatrix}, \tag{A.9a}$$

$$g(4) = H \begin{bmatrix} \frac{C_{13H}}{C_{33H}} & 0 & 0 \\ \frac{C_{23H}}{C_{33H}} & 0 & 0 \\ 0 & 0 & 0 \end{bmatrix}, \tag{A.9b}$$

$$g(5) = H \begin{bmatrix} C_{11H} - \frac{C_{13H}^2}{C_{33H}} & C_{12H} - \frac{C_{13H}C_{23H}}{C_{33b}} & 0 \\ C_{21H} - \frac{C_{13b}C_{23H}}{C_{33H}} & C_{22H} - \frac{C_{23H}^2}{C_{33H}} & 0 \\ 0 & 0 & C_{66H} \end{bmatrix}. \tag{A.9c}$$

Thus the sub-matrices of the stiffness for the horizontally fractured medium can be composed as follows:

$$C_{\mathrm{NN}\_\mathrm{VTI}} = \begin{bmatrix} \frac{C_{33H}}{1+S_{\mathrm{N}}C_{33H}} & 0 & 0 \\ 0 & \frac{C_{44H}}{1+S_{\mathrm{T}}C_{44H}} & 0 \\ 0 & 0 & \frac{C_{55H}}{1+S_{\mathrm{T}}C_{55H}} \end{bmatrix}, \tag{A.10a}$$

$$C_{\mathrm{TN\_VTI}} = \begin{bmatrix} \frac{C_{13H}}{1+S_N C_{33H}} & 0 & 0 \\ \frac{C_{23H}}{1+S_N C_{33H}} & 0 & 0 \\ 0 & 0 & 0 \end{bmatrix}, \tag{A.10b}$$

$$C_{\mathrm{TT\_VTI}} = \begin{bmatrix} C_{11H} - \frac{S_N C_{13H}^2}{1+S_N C_{33H}} & C_{12H} - \frac{S_N C_{13H} C_{23H}}{1+S_N C_{33H}} & 0 \\ C_{21H} - \frac{S_N C_{13H} C_{23H}}{1+S_N C_{33H}} & C_{22H} - \frac{S_N C_{23H}^2}{1+S_N C_{33H}} & 0 \\ 0 & 0 & C_{66H} \end{bmatrix}. \tag{A.10c}$$

Thus, the physical parameters of a horizontally fractured medium are

$$C_{\mathrm{VTI}} = \begin{bmatrix} (\lambda+2\mu) - \frac{S_N\lambda^2}{1+S_N(\lambda+2\mu)} & \lambda - \frac{S_N\lambda^2}{1+S_N(\lambda+2\mu)} & \frac{\lambda}{1+S_N(\lambda+2\mu)} & 0 & 0 & 0 \\ \lambda - \frac{S_N\lambda^2}{1+S_N(\lambda+2\mu)} & (\lambda+2\mu) - \frac{S_N\lambda^2}{1+S_N(\lambda+2\mu)} & \frac{\lambda}{1+S_N(\lambda+2\mu)} & 0 & 0 & 0 \\ \frac{\lambda}{1+S_N(\lambda+2\mu)} & \frac{\lambda}{1+S_N(\lambda+2\mu)} & \frac{\lambda+2\mu}{1+S_N(\lambda+2\mu)} & 0 & 0 & 0 \\ 0 & 0 & 0 & \frac{\mu}{1+S_T\mu} & 0 & 0 \\ 0 & 0 & 0 & 0 & \frac{\mu}{1+S_T\mu} & 0 \\ 0 & 0 & 0 & 0 & 0 & \mu \end{bmatrix}. \tag{A.11}$$

A.3 Moduli Calculation of the Vertically Fractured Medium

A rotation of 90° with respect to the y-axis can transform a VTI medium into a HTI medium by applying the orthogonal transformation matrix

$$\begin{pmatrix} \cos\theta & 0 & \sin\theta \\ 0 & 1 & 0 \\ -\sin\theta & 0 & \cos\theta \end{pmatrix}. \tag{A.12}$$

The corresponding Bond transformation matrix is

$$M = \begin{bmatrix} \cos^2\theta & 0 & \sin^2\theta & 0 & \sin 2\theta & 0 \\ 0 & 1 & 0 & 0 & 0 & 0 \\ \sin^2\theta & 0 & \cos^2\theta & 0 & -\sin 2\theta & 0 \\ 0 & 0 & 0 & \cos\theta & 0 & -\sin\theta \\ -\frac{1}{2}\sin 2\theta & 0 & \frac{1}{2}\sin 2\theta & 0 & \cos 2\theta & 0 \\ 0 & 0 & 0 & 0 & 0 & \cos\theta \end{bmatrix}. \tag{A.13}$$

For an HTI medium with a rotated symmetry axis (Winterstein 1990),

$$C_{\mathrm{HTI}} = M \cdot C_{\mathrm{VTI}} \cdot M^{\mathrm{T}}. \tag{A.14}$$

The HTI stiffness can be obtained from the stiffness of a VTI medium with a symmetric rotationally invariant system. The five independent parameters of VTI media are rotated as

$$\begin{bmatrix} C_{11_{\mathrm{VTI}}} & C_{12_{\mathrm{VTI}}} & C_{13_{\mathrm{VTI}}} & 0 & 0 & 0 \\ C_{12_{\mathrm{VTI}}} & C_{11_{\mathrm{VTI}}} & C_{13_{\mathrm{VTI}}} & 0 & 0 & 0 \\ C_{13_{\mathrm{VTI}}} & C_{13_{\mathrm{VTI}}} & C_{33_{\mathrm{VTI}}} & 0 & 0 & 0 \\ 0 & 0 & 0 & C_{55_{\mathrm{VTI}}} & 0 & 0 \\ 0 & 0 & 0 & 0 & C_{55_{\mathrm{VTI}}} & 0 \\ 0 & 0 & 0 & 0 & 0 & C_{66_{\mathrm{VTI}}} \end{bmatrix}$$
$$\rightarrow \begin{bmatrix} C_{33\_\mathrm{VTI}} & C_{13\_\mathrm{VTI}} & C_{13\_\mathrm{VTI}} & 0 & 0 & 0 \\ C_{13\_\mathrm{VTI}} & C_{11\_\mathrm{VTI}} & C_{12\_\mathrm{VTI}} & 0 & 0 & 0 \\ C_{13\_\mathrm{VTI}} & C_{12\_\mathrm{VTI}} & C_{11\_\mathrm{VTI}} & 0 & 0 & 0 \\ 0 & 0 & 0 & C_{66\_\mathrm{VTI}} & 0 & 0 \\ 0 & 0 & 0 & 0 & C_{55\_\mathrm{VTI}} & 0 \\ 0 & 0 & 0 & 0 & 0 & C_{55\_\mathrm{VTI}} \end{bmatrix}. \tag{A.15}$$

Thus the four sub-matrices of the rotated stiffness should be

$$C_{\mathrm{TT\_HTI}} = \begin{bmatrix} C_{33\_\mathrm{VTI}} & C_{13\_\mathrm{VTI}} & 0 \\ C_{13\_\mathrm{VTI}} & C_{11\_\mathrm{VTI}} & 0 \\ 0 & 0 & C_{55\_\mathrm{VTI}} \end{bmatrix}, \tag{A.16a}$$

$$C_{\mathrm{NN\_HTI}} = \begin{bmatrix} C_{11\_\mathrm{VTI}} & 0 & 0 \\ 0 & C_{66\mathrm{VTI}} & 0 \\ 0 & 0 & C_{55\mathrm{VTI}} \end{bmatrix}, \tag{A.16b}$$

$$C_{\mathrm{TN\_HTI}} = \begin{bmatrix} C_{13\_\mathrm{VTI}} & 0 & 0 \\ C_{12\_\mathrm{VTI}} & 0 & 0 \\ 0 & 0 & 0 \end{bmatrix}. \tag{A.16c}$$

Applying the Schoenberg and Muir (1989) calculation, the stiffness of the vertically fractured medium is

$$C_{\mathrm{HTI}} = \begin{bmatrix} \frac{\lambda+2\mu}{1+S_N(\lambda+2\mu)} & \frac{\lambda}{1+S_N(\lambda+2\mu)} & \frac{\lambda}{1+S_N(\lambda+2\mu)} & 0 & 0 & 0 \\ \frac{\lambda}{1+S_N(\lambda+2\mu)} & (\lambda+2\mu)-\frac{S_N\lambda^2}{1+S_N(\lambda+2\mu)} & \lambda-\frac{S_N\lambda^2}{1+S_N(\lambda+2\mu)} & 0 & 0 & 0 \\ \frac{\lambda}{1+S_N(\lambda+2\mu)} & \lambda-\frac{S_N\lambda^2}{1+S_N(\lambda+2\mu)} & (\lambda+2\mu)-\frac{S_N\lambda^2}{1+S_N(\lambda+2\mu)} & 0 & 0 & 0 \\ 0 & 0 & 0 & \mu & 0 & 0 \\ 0 & 0 & 0 & 0 & \frac{\mu}{1+S_T\mu} & 0 \\ 0 & 0 & 0 & 0 & 0 & \frac{\mu}{1+S_T\mu} \end{bmatrix}. \tag{A.17}$$

A.4 Moduli Calculation of the Orthogonally Fractured Medium

An orthorhombic medium has been considered as a realistic model and it mainly affects the permeability of the reservoir characterizations in geoscience. This medium is assumed to be a combination of the VTI medium and the HTI medium (Carcione 2012).

Take Eqs. (A.8a–A.8c) and (A.17) to implement the Schoenberg and Muir (1989) calculation:

$$g(3) = H/2\begin{bmatrix} \frac{C_{33}\_\mathrm{VTI}+C_{11}\_\mathrm{VTI}}{C_{33}\_\mathrm{VTI}C_{11}\_\mathrm{VTI}} & 0 & 0 \\ 0 & \frac{C_{55}\_\mathrm{VTI}+C_{66}\_\mathrm{VTI}}{C_{55}\_\mathrm{VTI}C_{66}\_\mathrm{VTI}} & 0 \\ 0 & 0 & \frac{2}{C_{55}\_\mathrm{VTI}} \end{bmatrix}, \tag{A.18a}$$

$$g(4) = H/2\begin{bmatrix} \frac{C_{11}\_\mathrm{VTI}C_{13}\_\mathrm{VTI}+C_{33}\_\mathrm{VTI}C_{13}\_\mathrm{VTI}}{C_{33}\_\mathrm{VTI}C_{11}\_\mathrm{VTI}} & 0 & 0 \\ \frac{C_{11}\_\mathrm{VTI}C_{13}\_\mathrm{VTI}+C_{33}\_\mathrm{VTI}C_{12}\_\mathrm{VTI}}{C_{33}\_\mathrm{VTI}C_{11}\_\mathrm{VTI}} & 0 & 0 \\ 0 & 0 & 0 \end{bmatrix}, \tag{A.18b}$$

$$g(5) = H/2, \tag{A.18c}$$

$$\begin{bmatrix} C_{11}\_\mathrm{VTI}+C_{33}\_\mathrm{VTI}-\frac{C_{11}\_\mathrm{VTI}C_{13}^2\_\mathrm{VTI}+C_{33}\_\mathrm{VTI}C_{13}^2\_\mathrm{VTI}}{C_{33}C_{11}} & C_{12}\_\mathrm{VTI}+C_{13}\_\mathrm{VTI}-\frac{C_{11}\_\mathrm{VTI}C_{13}^2\_\mathrm{VTI}+C_{33}\_\mathrm{VTI}C_{12}\_\mathrm{VTI}C_{13}\_\mathrm{VTI}}{C_{33}\_\mathrm{VTI}C_{11}\_\mathrm{VTI}} & 0 \\ C_{12}\_\mathrm{VTI}+C_{13}\_\mathrm{VTI}-\frac{C_{11}\_\mathrm{VTI}C_{13}^2\_\mathrm{VTI}+C_{13}\_\mathrm{VTI}C_{33}\_\mathrm{VTI}C_{12}\_\mathrm{VTI}}{C_{33}\_\mathrm{VTI}C_{11}\_\mathrm{VTI}} & 2C_{11}\_\mathrm{VTI}-\frac{C_{11}\_\mathrm{VTI}C_{13}^2\_\mathrm{VTI}+C_{33}\_\mathrm{VTI}C_{12}^2\_\mathrm{VTI}}{C_{33}\_\mathrm{VTI}C_{11}\_\mathrm{VTI}} & 0 \\ 0 & 0 & \frac{C_{55}\_\mathrm{VTI}+C_{66}\_\mathrm{VTI}}{C_{55}\_\mathrm{VTI}C_{66}\_\mathrm{VTI}} \end{bmatrix} \tag{A.18d}$$

Invert the orthorhombic group elements in Eq. (A.18a–A.18d) into the physical parameters by applying Eq. (A.7)

$$C_{TT\_ORTH} = \begin{bmatrix} \frac{C_{11\_VTI} + C_{33\_VTI}}{2} & \frac{C_{12\_VTI} + C_{13\_VTI}}{2} & 0 \\ \frac{C_{12\_VTI} + C_{13\_VTI}}{2} & C_{11\_VTI} - \frac{(C_{12\_VTI} - C_{13\_VTI})^2}{2(C_{11\_VTI} + C_{33\_VTI})} & 0 \\ 0 & 0 & \frac{C_{55\_VTI} + C_{66\_VTI}}{2} \end{bmatrix}, \tag{A.19a}$$

$$C_{TN\_ORTH} = \begin{bmatrix} C_{13\_VTI} & 0 & 0 \\ \frac{C_{11\_VTI}C_{13\_VTI} + C_{12\_VTI}C_{33\_VTI}}{C_{11\_VTI} + C_{13\_VTI}} & 0 & 0 \\ 0 & 0 & 0 \end{bmatrix}, \tag{A.19b}$$

$$C_{NN\_ORTH} = \begin{bmatrix} \frac{2C_{33\_VTI}C_{11\_VTI}}{C_{33\_VTI} + C_{11\_VTI}} & 0 & 0 \\ 0 & \frac{2C_{55\_VTI}C_{66\_VTI}}{C_{55\_VTI} + C_{66\_VTI}} & 0 \\ 0 & 0 & C_{55\_VTI} \end{bmatrix}, \tag{A.18c}$$

Then the stiffness of the orthogonally fracture media are

$$C_{ORTH} = \begin{bmatrix} \frac{C_{11\_VTI} + C_{33\_VTI}}{2} & \frac{C_{12\_VTI} + C_{13\_VTI}}{2} & C_{13\_VTI} & 0 & 0 & 0 \\ \frac{C_{12\_VTI} + C_{13\_VTI}}{2} & C_{11\_VTI} - \frac{(C_{12\_VTI} - C_{13\_VTI})^2}{2(C_{11} + C_{33})} & \frac{C_{11\_VTI}C_{13\_VTI} + C_{12\_VTI}C_{33\_VTI}}{C_{11\_VTI} + C_{13\_VTI}} & 0 & 0 & 0 \\ C_{13\_VTI} & \frac{C_{11\_VTI}C_{13\_VTI} + C_{12\_VTI}C_{33\_VTI}}{C_{11\_VTI} + C_{13\_VTI}} & \frac{2C_{33\_VTI}C_{11\_VTI}}{C_{33\_VTI} + C_{11\_VTI}} & 0 & 0 & 0 \\ 0 & 0 & 0 & \frac{2C_{55\_VTI}(C_{11\_VTI} - C_{12\_VTI})}{2C_{55\_VTI} + C_{11\_VTI} - C_{12\_VTI}} & 0 & 0 \\ 0 & 0 & 0 & 0 & C_{55\_VTI} & 0 \\ 0 & 0 & 0 & 0 & 0 & \frac{1}{4}(2C_{55\_VTI} + C_{11\_VTI}) \end{bmatrix}, \tag{A.20}$$

where $C_{66} = \frac{1}{2}(C_{11} - C_{12})$.

Appendix B
Numerical Operators

B.1 Average Operator

Define an operator $\mathbb{A}$ at point x as follows (see Fig. B.1)

$$\mathbb{A}f(x) = \frac{1}{2}\left[f\left(x + \frac{h}{2}\right) + f\left(x - \frac{h}{2}\right)\right]. \tag{B.1}$$

$\mathbb{A}$ is called the average operator.

B.2 Difference Operator

The finite difference operators can be defined in terms of forward, central, and backward operators, and their consecutive expressions in the first order are

$$\left.\frac{\partial f}{\partial x}\right|_{(x+\frac{h}{2})} = \frac{f(x+h) - f(x)}{h}, \tag{B.2a}$$

$$\left.\frac{\partial f}{\partial x}\right|_{x} = \frac{f(x+h/2) - f(x-h/2)}{h}, \tag{B.2b}$$

$$\left.\frac{\partial f}{\partial x}\right|_{(x-\frac{h}{2})} = \frac{f(x) - f(x-h)}{h}. \tag{B.2c}$$

Note that Eq. (B.2a–B.2c) has different center points, i.e., $x + \frac{1}{2}h$, x, and $x - \frac{1}{2}h$. Consider a second-order derivative with the center point at x. We have

$$\left.\frac{\partial^2 f}{\partial x^2}\right|_{x} = \frac{\frac{\partial}{\partial x}f(x+h/2) - \frac{\partial}{\partial x}f(x-h/2)}{h} = \frac{f(x+h) - 2f(x) + f(x-h)}{h^2}, \tag{B.3}$$

Equation (B.3) is the second-order finite-difference operator.

© Springer Nature Singapore Pte Ltd. 2018

X. Cui et al., *Seismic Forward Modeling of Fractures and Fractured Medium Inversion*, Springer Geophysics, DOI 10.1007/978-981-10-3584-5

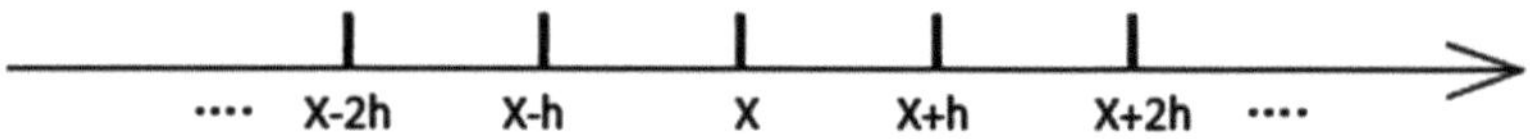

Fig. B.1 Delineation steps for numerical operator

B.3 Accuracy of Operator

In an application of the finite-difference method to the PDE wave equation in seismology one needs to be aware of the accuracy and numerical stability issues to avoid numerical grid dispersion and oscillation problems. The numerical grid dispersion is caused by truncation of higher order terms in Taylor series expansion. So the PDE solution is inaccurate for the derivative, which leads propagation velocities of seismic waves to be frequency dependent.

Expanding a function f in a Taylor series, one obtains $f(x+h)$ as

$$f(x+h) = f(x) + h\frac{\partial f}{\partial x}\bigg|_x + \frac{h^2}{2!}\left(\frac{\partial^2 f}{\partial x^2}\right)\bigg|_x + \frac{h^3}{3!}\left(\frac{\partial^3 f}{\partial x^3}\right)\bigg|_x + \cdots. \tag{B.4a}$$

The forward difference operator is

$$\frac{f(x+h) - f(x)}{h} = \frac{\partial f}{\partial x}\bigg|_x + \text{err}^+, \tag{B.4b}$$

$$\text{err}^+ = \frac{h}{2!}\left(\frac{\partial^2 f}{\partial x}\right)\bigg|_x + \frac{h^2}{3!}\left(\frac{\partial^3 f}{\partial x^3}\right)\bigg|_x \approx O(h). \tag{B.4c}$$

Expanding a function f in a Taylor series, one obtains $f(x-h)$ as

$$f(x-h) = f(x) + (-h)\frac{\partial f}{\partial x}\bigg|_x + \frac{(-h)^2}{2!}\left(\frac{\partial^2 f}{\partial x^2}\right)\bigg|_x + \frac{(-h)^3}{3!}\left(\frac{\partial^3 f}{\partial x^3}\right)\bigg|_x + \cdots. \tag{B.5a}$$

The backward difference operator is

$$\frac{f(x) - f(x-h)}{h} = \frac{\partial f}{\partial x}\bigg|_x + \text{err}^-, \tag{B.5b}$$

$$\text{err}^- = \frac{(-h)}{2!}\left(\frac{\partial^2 f}{\partial x^2}\right)\bigg|_x + \frac{(-h)^2}{3!}\left(\frac{\partial^3 f}{\partial x^3}\right)\bigg|_x + \cdots \approx O(h). \tag{B.5c}$$

We can obtain $f(x+h) - f(x-h)$ by subtracting Eq. (B.4a) from Eq. (B.5a) as

$$f(x+h) - f(x-h) = 2h\frac{\partial f}{\partial x}\bigg|_x + 2\frac{h^3}{3!}\left(\frac{\partial^3 f}{\partial x^3}\right)\bigg|_x + \cdots. \tag{B.6}$$

Hence, the central difference operator to the first order derivative is

$$\frac{f(x+h) - f(x-h)}{2h} = \frac{\partial f}{\partial x}\bigg|_x + \text{err}, \tag{B.7a}$$

$$\text{err} = \frac{h^2}{3!}\left(\frac{\partial^3 f}{\partial x^3}\right)\bigg|_x + \cdots \approx O(h^2). \tag{B.7b}$$

We can obtain $f(x+h) + f(x-h)$ by summing Eqs. (C.4a) and (C.5)

$$f(x+h) + f(x-h) = 2f(x) + 2\frac{h^2}{2!}\left(\frac{\partial^2 f}{\partial x^2}\right)\bigg|_x + \cdots. \tag{B.8a}$$

So the central difference operator for the second derivative is

$$\frac{f(x+h) - 2f(x) + f(x-h)}{h^2} = \frac{\partial^2 f}{\partial x^2}\bigg|_x + \text{err}, \tag{B.8b}$$

$$\text{err} = \frac{h^2}{4!}\left(\frac{\partial^4 f}{\partial x^4}\right)\bigg|_x + \cdots \approx O(h^2). \tag{B.8c}$$

err, err^+ and err^- are the so-called truncation errors. The lowest power of h in the truncation error is the order of accuracy of the finite-difference approximation. Thus, in the first order derivative, the forward and backward difference formulas had an error of the first order $O(h)$, while the central difference formula yields an error of the second-order $O(h^2)$.

Appendix C
PP Reflection Coefficients for the Fractured Medium

C.1 Waves at a Nonwelded Contact Fracture Interface

Consider fractured media in which the incident P-wave, and reflected PP- and PS-waves are in the upper medium, and the transmission waves PP and PS are in the lower medium. The fractured medium is composed of a fracture and an anisotropic host medium. The displacement $\boldsymbol{u}$ and the stresses $\boldsymbol{\sigma}$ of a wave are

$$\boldsymbol{u} = \mathrm{A}e^{-i\omega(t-\boldsymbol{s}\cdot\boldsymbol{x})}\boldsymbol{d}, \tag{C.1}$$

$$\sigma_{xz} = \mu\left(\frac{\partial u_x}{\partial z} + \frac{\partial u_z}{\partial x}\right) = i\omega \mathrm{A}e^{-i\omega(t-\boldsymbol{s}\cdot\boldsymbol{x})}\mu[s_x d_z + s_z d_x], \tag{C.2}$$

$$\sigma_{zz} = \lambda\frac{\partial u_x}{\partial x} + (\lambda+2\mu)\left(\frac{\partial u_z}{\partial z}\right) = i\omega \mathrm{A}e^{-i\omega(t-\boldsymbol{s}\cdot\boldsymbol{x})}[\lambda s_x d_x + (\lambda+2\mu)s_z d_z], \tag{C.3}$$

where the amplitude A, slowness $\boldsymbol{s}$, and polarization vector $\boldsymbol{d}$ are different for each wave. The sign of the polarization vector $\boldsymbol{d}$ follows the conventional notation that the component in the x-axis is positive $d_x > 0$.

Table C.1 trigonometrically provides that the plane wave of amplitude, slowness, and polarization expressions in x, z-component should be obtained.

The boundary conditions of viscous fracture interface are that $\sigma(xz)$ and $\sigma(zz)$ are continuous across the interface, while $u(x)$, $u(z)$ are discontinuous across the interface. The displacement difference are linearly proportional to the stresses. Thus, the displacements $\boldsymbol{u}$ of all plane wave are constrained by the nonwelded boundary conditions as

© Springer Nature Singapore Pte Ltd. 2018

X. Cui et al., *Seismic Forward Modeling of Fractures and Fractured Medium Inversion*, Springer Geophysics, DOI 10.1007/978-981-10-3584-5

Table C.1 Amplitude, slowness, and polarization for plane wave at the VTI interface

Waves	Amplitude	s_x	s_z	d_x	d_z
$P^{\backprime}$	$P_1^{\backprime}$	$\frac{\sin\theta_1}{\alpha_1}$	$\frac{\cos\theta_1}{\alpha_1}$	$l_{\theta 1}\sin\theta_1$	$m_{\theta 1}\cos\theta_1$
$P^{\backprime}P'$	P_1'	$\frac{\sin\theta_1}{\alpha_1}$	$-\frac{\cos\theta_1}{\alpha_1}$	$l_{\theta 1}\sin\theta_1$	$-m_{\theta 1}\cos\theta_1$
$P^{\backprime}S'$	S_1'	$\frac{\sin\vartheta_1}{\beta_1}$	$-\frac{\cos\vartheta_1}{\beta_1}$	$m_{\vartheta 1}\cos\vartheta_1$	$l_{\vartheta 1}\sin\vartheta_1$
$P^{`}P^{`}$	$P_2^{`}$	$\frac{\sin\theta_2}{\alpha_2}$	$\frac{\cos\theta_2}{\alpha_2}$	$l_{\theta 2}\sin\theta_2$	$m_{\theta 2}\cos\theta_2$
$P^{`}S^{`}$	$S_2^{`}$	$\frac{\sin\vartheta_2}{\beta_2}$	$\frac{\cos\vartheta_2}{\beta_2}$	$m_{\vartheta 2}\cos\vartheta_2$	$-l_{\vartheta 2}\sin\vartheta_2$

$$
\begin{aligned}
P_1^{\backprime} l_{\theta 1}\sin\theta_1 = & -P_1' l_{\theta 1}\sin\theta_1 - S_1' m_{\vartheta 1}\cos\vartheta_1 \\
& + P_2^{\backprime}\left(l_{\theta 2}\sin\theta_2 - \frac{i\omega S_T}{1 - i\omega v_x S_T}(xx)_2\cos\theta_2\right) \\
& + S_2^{\backprime}\left(m_{\vartheta 2}\cos\vartheta_2 - \frac{i\omega S_T}{1 - i\omega v_x S_T}\beta_2 (rr)_2\right),
\end{aligned}
\tag{C.4a}
$$

$$
\begin{aligned}
P_1^{\backprime} m_{\theta 1}\cos\theta_1 = & \, P_1' m_{\theta 1}\cos\theta_1 - S_1' l_{\vartheta 1}\sin\vartheta_1 \\
& + P_2^{\backprime}\left(m_{\theta 2}\cos\theta_2 - i\omega S_N \alpha_2 (yy)_2\right) \\
& + S_2^{\backprime}\left(-l_{\vartheta 2}\sin\vartheta_2 + i\omega S_N (kk)_2\cos\vartheta_2\right),
\end{aligned}
\tag{C.4b}
$$

The stresses $\boldsymbol{\sigma}$ of the nonwelded boundary condition are

$$
P_1^{\backprime}(xx)_1\cos\theta_1 = P_1'(xx)_1\cos\theta_1 + S_1'(rr)_1\beta_1 + P_2^{\backprime}(xx)_2\cos\theta_2 + S_2^{\backprime}(rr)_2\beta_2, \tag{C.4c}
$$

$$
P_1^{\backprime}(yy)_1\alpha_1 = -P_1'(yy)_1\alpha_1 + S_1'(kk)_1\cos\vartheta_1 + P_2^{\backprime}(yy)_2\alpha_2 - S_2^{\backprime}(kk)_2\cos\vartheta_2, \tag{C.4d}
$$

where $(xx)_N$, $(rr)_N$, $(yy)_N$, and $(kk)_N$ are defined in Eq. (4.10e–h). $n = 1, 2$. Equations (C.4a–C.4d) express the relationships between all waves for the fractured medium that is composed of a fracture and an anisotropic host medium.

C.2 Exact Solution of PP Reflection Coefficients for Fractured Medium

If we let $l_{\theta n} = m_{\theta n} = l_{\vartheta n} = m_{\vartheta n} = 1$, then the anisotropic host mediumchanges to an isotropic medium. Equations (C.4a–C.4d) can be rearranged into Eq. (4.9). According to Cramer's rule,

$$
\begin{aligned}
P_1'P_1' = R(\theta) &= \frac{\det(M_{\mathrm{pp}})}{\det(M)} \\
&= \frac{(bq_{\alpha1} - cq_{\alpha2})F - (a + dq_{\alpha1}q_{\beta2})\mathrm{HP}^2 + (\rho_2 q_{\alpha2}L_1 - \rho_1 q_{\alpha1}K_2)\frac{i\omega S_{\mathrm{T}}}{1-i\omega v_x S_{\mathrm{T}}} + (\rho_2 q_{\beta2}L_1 + \rho_1 q_{\beta1}K_2)i\omega S_{\mathrm{N}} + L_1K_2\omega^2\frac{S_{\mathrm{T}}}{1-i\omega v_x S_{\mathrm{T}}}S_{\mathrm{N}}}{(\mathrm{EF} + \mathrm{GHP}^2) - (\rho_2 q_{\alpha2}K_1 + \rho_1 q_{\alpha1}K_2)\frac{i\omega S_{\mathrm{T}}}{1-i\omega v_x S_{\mathrm{T}}} - (\rho_2 q_{\beta2}L_1 + \rho_1 q_{\beta1}K_2)i\omega S_{\mathrm{N}} - K_1K_2\omega^2\frac{S_{\mathrm{T}}}{1-i\omega v_x S_{\mathrm{T}}}S_{\mathrm{N}}} \\
&= \frac{N}{D} = \frac{N_{\mathrm{w}} + N_{\mathrm{non\_w}}}{D_{\mathrm{w}} + D_{\mathrm{non\_w}}} = \frac{N_{\mathrm{w}} + N^{\mathrm{T}}_{\mathrm{non\_w}}\frac{i\omega S_{\mathrm{T}}}{1-i\omega v_x S_{\mathrm{T}}} + N^{\mathrm{N}}_{\mathrm{non\_w}}i\omega S_{\mathrm{N}} + N^{\mathrm{TN}}_{\mathrm{non\_w}}\omega^2\frac{S_{\mathrm{T}}}{1-i\omega v_x S_{\mathrm{T}}}S_{\mathrm{N}}}{D_{\mathrm{w}} + D^{\mathrm{T}}_{\mathrm{non\_w}}\frac{i\omega S_{\mathrm{T}}}{1-i\omega v_x S_{\mathrm{T}}} + D^{\mathrm{N}}_{\mathrm{non\_w}}i\omega S_{\mathrm{N}} + D^{\mathrm{TN}}_{\mathrm{non\_w}}\omega^2\frac{S_{\mathrm{T}}}{1-i\omega v_x S_{\mathrm{T}}}S_{\mathrm{N}}} \\
&= \frac{N_{\mathrm{w}}}{D_{\mathrm{w}}} + \frac{D^{\mathrm{T}}_{\mathrm{non\_w}}R_{\mathrm{w}} + N^{\mathrm{T}}_{\mathrm{non\_w}}}{D}\frac{i\omega S_{\mathrm{T}}}{1 - i\omega v_x S_{\mathrm{T}}} + \frac{D^{\mathrm{N}}_{\mathrm{non\_w}}R_{\mathrm{w}} + N^{\mathrm{N}}_{\mathrm{non\_w}}}{D}i\omega S_{\mathrm{N}} + \frac{D^{\mathrm{TN}}_{\mathrm{non\_w}}R_{\mathrm{w}} + N^{\mathrm{TN}}_{\mathrm{non\_w}}}{D}\omega^2\frac{S_{\mathrm{T}}}{1 - i\omega v_x S_{\mathrm{T}}}S_{\mathrm{N}},
\end{aligned}
\tag{C.5}
$$

where a, b, c, d, $q_{\alpha n}$, $q_{\beta n}$, E, F, G, H, L and K have definitions similar to those given by Aki and Richards (1980), or Chaisri and Krebes (2000).

$$
\begin{aligned}
a = x_2 - x_1, \quad b = r_2 + x_1P, \quad c = r_1 + x_2P, \\
d = 2(\rho_2\beta_2^2 - \rho_1\beta_1^2) = \frac{1}{P}(x_2 - x_1),
\end{aligned}
\tag{C.6a}
$$

$$
q_{\alpha n} = \frac{\cos\theta_{\mathrm{N}}}{\alpha_{\mathrm{N}}}, \quad q_{\beta n} = \frac{\cos\vartheta_{\mathrm{N}}}{\beta_{\mathrm{N}}},
\tag{C.6b}
$$

$$
\begin{aligned}
E = (bq_{\alpha1} + cq_{\alpha2}), \quad F = (bq_{\beta1} + cq_{\beta2}), \quad G = (a - dq_{\alpha1}q_{\beta2}), \\
H = (a - dq_{\alpha2}q_{\beta1}),
\end{aligned}
\tag{C.7a}
$$

$$
K_{\mathrm{N}} = (r_{\mathrm{N}}^2 + x_{\mathrm{N}}^2)q_{\alpha n}q_{\beta n}, \quad L_{\mathrm{N}} = (r_{\mathrm{N}}^2 - x_{\mathrm{N}}^2)q_{\alpha n}q_{\beta n}.
\tag{C.7b}
$$

C.3 Approximate PP AVO Equation for Fractured Medium

Note that the exact solution for the PP reflection coefficients in Eq. (C.5) includes a perfect welded contact part and imperfectly welded contact part. For the imperfectly welded contact part, $\frac{R_{\mathrm{w}}}{D_{\mathrm{w}}} \ll 1$, and if we only take the first order in the tangential term S_{T}, and the normal term S_{N}, Eq. (C.5) is approximated as (Chaisri 2002)

$$
\begin{aligned}
R(\theta) &\approx \frac{\mathrm{N_w}}{\mathrm{D_w}} + \frac{1}{\mathrm{D_w}}(\rho_2 q_{\alpha2}L_1 - \rho_1 q_{\alpha1}K_2)\frac{i\omega S_{\mathrm{T}}}{1 - i\omega v_x S_{\mathrm{T}}} + (\rho_2 q_{\beta2}L_1 + \rho_1 q_{\alpha1}K_2)i\omega S_{\mathrm{N}} \\
&\approx \frac{\mathrm{N_w}}{\mathrm{D_w}} + \frac{\mathrm{N}^{\mathrm{T}}_{\mathrm{non\_w}}}{\mathrm{D_w}}\frac{i\omega S_{\mathrm{T}}}{1 - i\omega v_x S_{\mathrm{T}}} + \frac{\mathrm{N}^{\mathrm{N}}_{\mathrm{non\_w}}}{\mathrm{D_w}}i\omega S_{\mathrm{N}} \approx R_{\mathrm{w}} + \frac{i\omega S_{\mathrm{T}}}{1 - i\omega v_x S_{\mathrm{T}}}R_{\mathrm{non\_}}\mathrm{w}^{\mathrm{T}} + i\omega S_{\mathrm{N}}R^{\mathrm{N}}_{\mathrm{non\_w}}.
\end{aligned}
\tag{C.8}
$$

Extend and rearrange the terms of N_{w}, D_{w} in the orders of the ray parameter P,

$$\begin{aligned}N_{\mathrm{w}} &= (\rho_2 q_{\alpha 1} - \rho_1 q_{\alpha 2})(\rho_2 q_{\beta 1} + \rho_1 q_{\beta 2})\\ &\quad + \left[-4(\Delta\mu)(\rho_2 q_{\alpha 1} q_{\beta 1} + \rho_1 q_{\alpha 2} q_{\beta 2}) - \Delta\rho^2 + 4(\Delta\mu)^2 q_{\alpha 1} q_{\alpha 2} q_{\beta 1} q_{\beta 2}\right] P^2\\ &\quad + \left[4(\Delta\mu)^2 (q_{\alpha 1} q_{\beta 1} - q_{\alpha 2} q_{\beta 2})\right] P^4 - 4(\Delta\mu)^2 P^6\\ &= E_{\mathrm{w}} + F_{\mathrm{w}} P^2 + G_{\mathrm{w}} P^4 - H_{\mathrm{w}} P^6,\end{aligned} \tag{C.9a}$$

$$\begin{aligned}D_{\mathrm{w}} &= (\rho_2 q_{\alpha 1} + \rho_1 q_{\alpha 2})(\rho_2 q_{\beta 1} + \rho_1 q_{\beta 2})\\ &\quad + \left[-4(\Delta\mu)(\rho_2 q_{\alpha 1} q_{\beta 1} - \rho_1 q_{\alpha 2} q_{\beta 2}) + \Delta\rho^2 + 4(\Delta\mu)^2 q_{\alpha 1} q_{\alpha 2} q_{\beta 1} q_{\beta 2}\right] P^2\\ &\quad + \left[4(\Delta\mu)^2 (q_{\alpha 1} q_{\beta 1} + q_{\alpha 2} q_{\beta 2})\right] P^4 + 4(\Delta\mu)^2 P^6\\ &= A_{\mathrm{w}} + B_{\mathrm{w}} P^2 + C_{\mathrm{w}} P^4 + I_{\mathrm{w}} P^6.\end{aligned} \tag{C.9b}$$

So, the reflection coefficients of the perfectly welded part in Eq. (C.8) can be approximated to second order in P as

$$\frac{N_{\mathrm{w}}}{D_{\mathrm{w}}} \approx \frac{E_{\mathrm{w}}}{A_{\mathrm{w}}} + \left(\frac{F_{\mathrm{w}}}{A_{\mathrm{w}}} - \frac{B_{\mathrm{w}} E_{\mathrm{w}}}{A_{\mathrm{w}}^2}\right) P^2. \tag{C.10}$$

Apply the same approximation as Aki and Richards (1980) in which are $\Delta\theta = \theta_2 - \theta_1 = \tan\theta \frac{\Delta\alpha}{\alpha}$, $\Delta\vartheta = \vartheta_2 - \vartheta_1 = \tan\vartheta \frac{\Delta\beta}{\beta}$. $\alpha = \frac{\alpha_2 + \alpha_1}{2}$, $\Delta\alpha = \alpha_2 - \alpha_1$. $\beta = \frac{\beta_2 + \beta_1}{2}$, $\Delta\beta = \beta_2 - \beta_1$. $\rho = \frac{\rho_2 + \rho_1}{2}$, $\Delta\rho = \rho_2 - \rho_1$. to Eq. (C.10). Then

$$\begin{aligned}\frac{E_{\mathrm{w}}}{A_{\mathrm{w}}} &= \frac{(\rho_2 q_{\alpha 1} - \rho_1 q_{\alpha 2})}{(\rho_2 q_{\alpha 1} + \rho_1 q_{\alpha 2})} = R_f\\ &= \frac{\left(\rho + \frac{\Delta\rho}{2}\right)\left(\alpha + \frac{\Delta\alpha}{2}\right)\cos\left(i - \frac{\Delta i}{2}\right) - \left(\rho - \frac{\Delta\rho}{2}\right)\left(\alpha - \frac{\Delta\alpha}{2}\right)\cos\left(i + \frac{\Delta i}{2}\right)}{\left(\rho + \frac{\Delta\rho}{2}\right)\left(\alpha + \frac{\Delta\alpha}{2}\right)\cos\left(i - \frac{\Delta i}{2}\right) + \left(\rho - \frac{\Delta\rho}{2}\right)\left(\alpha - \frac{\Delta\alpha}{2}\right)\cos\left(i + \frac{\Delta i}{2}\right)}\\ &= \frac{\frac{\Delta\rho}{\rho} + \frac{\Delta\alpha}{\alpha} + \left(2 + \frac{1}{2}\frac{\Delta\rho}{\rho}\frac{\Delta\alpha}{\alpha}\right)\tan i \tan\frac{\Delta i}{2}}{2 + \frac{1}{2}\frac{\Delta\rho}{\rho}\frac{\Delta\alpha}{\alpha} + \left(\frac{\Delta\rho}{\rho} + \frac{\Delta\alpha}{\alpha}\right)\tan i \tan\frac{\Delta i}{2}}\\ &= \frac{\frac{\Delta\rho}{\rho} + \frac{\Delta\alpha}{\alpha} + \left(2 + \frac{1}{2}\frac{\Delta\rho}{\rho}\frac{\Delta\alpha}{\alpha}\right)\tan i \tan\frac{\Delta i}{2}}{2 + \frac{1}{2}\frac{\Delta\rho}{\rho}\frac{\Delta\alpha}{\alpha} + \left(\frac{\Delta\rho}{\rho} + \frac{\Delta\alpha}{\alpha}\right)\tan i \tan\frac{\Delta i}{2}}\\ &\approx \frac{1}{2}\left(\frac{\Delta\rho}{\rho} + \frac{\Delta\alpha}{\alpha}\sec\theta^2\right),\end{aligned} \tag{C.11a}$$

$$\frac{F_{\mathrm{w}}}{A_{\mathrm{w}}}-\frac{B_{\mathrm{w}}E_{\mathrm{w}}}{A_{\mathrm{w}}^{2}}\approx-\frac{1}{A_{\mathrm{w}}}\big[4(\Delta\mu)\big(\rho_2 q_{\alpha1}q_{\beta1}+\rho_1 q_{\alpha2}q_{\beta2}\big)$$
$$-4(\Delta\mu)R_f\big(\rho_2 q_{\alpha1}q_{\beta1}-\rho_1 q_{\alpha2}q_{\beta2}\big)-4(\Delta\mu)^2 q_{\alpha1}q_{\alpha2}q_{\beta1}q_{\beta2}\big] \quad \text{(C.11b)}$$
$$\approx-2\frac{(\Delta\mu)}{\rho}+\big(1-R_f\big)\left(\frac{(\Delta\mu)}{\rho}\right)^2 q_\alpha q_\beta,$$

in which, $(\Delta\mu)=\rho_2\beta_2^2-\rho_1\beta_1^2$, $q_\alpha=\frac{q_{\alpha1}+q_{\alpha2}}{2}$, $q_\beta=\frac{q_{\beta1}+q_{\beta2}}{2}$, $\Delta q_\alpha=q_{\alpha2}-q_{\alpha1}$, $\Delta q_\beta=q_{\beta2}-q_{\beta1}$. Then,

$$R_{\mathrm{w}}\approx R_f-(2\frac{(\Delta\mu)}{\rho}-(1-\mathrm{R_f})\left(\frac{(\Delta\mu)}{\rho}\right)^2 q_\alpha q_\beta\Big)P^2$$
$$\approx R_f-2\left(2\beta\Delta\beta+\frac{\Delta\rho}{\rho}\beta^2\right)\frac{\sin^2\theta}{\alpha^2} \quad \text{(C.12)}$$
$$\approx\frac{1}{2\cos^2\theta}r_\alpha-4\left(\frac{\beta}{\alpha}\right)^2\sin^2\theta r_\beta+\frac{1}{2}\left(1-4\left(\frac{\beta}{\alpha}\right)^2\sin^2\theta\right)r_\rho.$$

Similarly, the tangential term S_{T} of the imperfectly welded part is

$$\frac{N_{\mathrm{non\_w}}^{\mathrm{T}}}{D_{\mathrm{w}}}=\frac{E_{\mathrm{non\_w}}^{\mathrm{T}}}{A_{\mathrm{w}}}+\left(\frac{F_{\mathrm{non\_w}}^{\mathrm{T}}}{A_{\mathrm{w}}}-\frac{B_{\mathrm{w}}E_{\mathrm{non\_w}}^{\mathrm{T}}}{A_{\mathrm{w}}^{2}}\right)P^2, \quad \text{(C.13a)}$$

where

$$\frac{E_{\mathrm{non_w}}^{\mathrm{T}}}{A_{\mathrm{w}}}=\frac{\rho_1^2\rho_2 q_{\alpha2}-\rho_1\rho_2^2 q_{\alpha1}}{(\rho_2 \mathrm{q}_{\alpha1}+\rho_1\mathrm{q}_{\alpha2})(\rho_2\mathrm{q}_{\beta1}+\rho_1\mathrm{q}_{\beta2})}$$
$$\approx-R_f\frac{\rho_1\rho_2\beta_1\beta_2}{\rho_2\beta_2\cos j_1+\rho_1\beta_1\cos j_2}$$
$$\approx-R_f\frac{\rho\beta}{[\left(1+\frac{\Delta\rho}{2\rho}+\frac{\Delta\beta}{2\beta}\right)\left(1+\frac{1}{2}\frac{\Delta\beta}{\beta}\tan^2 j)]\right)+\left(1-\frac{\Delta\rho}{2\rho}-\frac{\Delta\beta}{2\beta}\right)\left(1-\frac{1}{2}\frac{\Delta\beta}{\beta}\tan^2 j)]\right)](\cos j\cos\frac{\Delta j}{2})}$$
$$\approx-\frac{1}{2}R_f\rho\beta\sec\vartheta=-\frac{1}{4}\rho\beta\sec\vartheta\frac{\Delta\rho}{\rho}-\frac{1}{4}\rho\beta\sec\vartheta\sec\theta^2\frac{\Delta\alpha}{\alpha},$$
$$\text{(C.13b)}$$

$$\frac{F_{\mathrm{non_w}}^{\mathrm{T}}}{A_{\mathrm{w}}}\approx\left(\frac{4\rho_1\rho_2(\mu_2 q_{\alpha1}-\mu_1 q_{\alpha2})}{\mathrm{A_w}}-\frac{4\rho_1 q_{\alpha1}q_{\alpha2}\big(\mu_2^2 q_{\beta2}+\mu_1^2 q_{\beta1}\big)}{\mathrm{A_w}}\right)$$
$$\approx\frac{\mu}{\mathrm{q}_\beta}\left(\frac{\Delta\mu}{\mu}-\frac{\Delta q_\alpha}{q_\alpha}\right)+2\frac{1}{\mathrm{\rho}}\mu^2 q_\alpha\left[1+\frac{1}{4}\left(\frac{\Delta\mu}{\mu}\right)^2+\frac{1}{2}\left(\frac{\Delta q_\beta}{q_\beta}\right)\left(\frac{\Delta\mu}{\mu}\right)\right] \quad \text{(C.13c)}$$
$$\approx\frac{\mu}{q_\beta}\left(2\frac{\Delta\beta}{\beta}+\frac{\Delta\rho}{\rho}+\frac{\Delta\alpha}{\alpha}\right)P^2+2\frac{1}{\rho}\mu^2 q_\alpha P^2,$$

Derivation of Eq. (C.13a, C.13b) used the following approximations:

$$\cos\frac{\Delta\vartheta}{2} \approx 1, (\cos\vartheta) = 1 - \frac{1}{2}\left(\frac{\beta}{\alpha}\right)^2 \sin\theta^2, \tag{C.14a}$$

$$(\cos\vartheta) = 1 - \frac{1}{2}\left(\frac{\beta}{\alpha}\right)^2 \sin\theta^2, \tag{C.14b}$$

$$\frac{\Delta q_\beta}{q_\beta} \approx -\frac{\Delta\beta}{\beta} - \frac{\Delta\beta}{\beta}\beta_1\beta_2 P^2, \quad \frac{\Delta q_\alpha}{q_\alpha} \approx -\frac{\Delta\alpha}{\alpha} - \frac{\Delta\alpha}{\alpha}\alpha_1\alpha_2 P^2. \tag{C.14c}$$

Thus, the tangential term S_{T} in the nonwelded part is finally approximated as

$$\begin{aligned}\frac{\mathrm{N}^{\mathrm{T}}_{\mathrm{non_w}}}{\mathrm{D_w}} &\approx 2\frac{1}{\rho}\mu^2 q_\alpha P^2 + \left(\frac{\mu}{q_\beta}P^2 - \frac{1}{4}\rho\beta\sec\vartheta\right)\frac{\Delta\rho}{\rho} \\ &+ \left(\frac{\mu}{q_\beta}P^2 - \frac{1}{4}\rho\beta\sec\vartheta\sec\theta^2\right)\frac{\Delta\alpha}{\alpha} + 2\frac{\mu}{q_\beta}P^2\frac{\Delta\beta}{\beta}.\end{aligned} \tag{C.15}$$

Similarly, the normal term in the nonwelded part can be rearranged as

$$\begin{aligned}\frac{\mathrm{N}^{\mathrm{N}}_{\mathrm{non_w}}}{\mathrm{D_w}} &\approx \left(\frac{\mathrm{E}^{\mathrm{N}}_{\mathrm{non_w}} + \mathrm{F}^{\mathrm{N}}_{non_w}P^2}{\mathrm{D_w}}\right) \\ &\approx \frac{\mathrm{E}^{\mathrm{N}}_{\mathrm{non\_w}}}{\mathrm{A_w}} + \left(\frac{\mathrm{F}^{\mathrm{N}}_{\mathrm{non-w}}}{\mathrm{A_w}} - \frac{\mathrm{B_w}\mathrm{E}^{\mathrm{N}}_{\mathrm{non-w}}}{\mathrm{A_w^2}}\right)P^2,\end{aligned} \tag{C.16a}$$

$$\frac{\mathrm{E}^{\mathrm{N}}_{\mathrm{non\_w}}}{\mathrm{A_w}} = \frac{\rho_1^2\rho_2 q_{\beta2} + \rho_1\rho_2^2 q_{\beta1}}{(\rho_2 q_{\alpha1} + \rho_1 q_{\alpha2})(\rho_2 q_{\beta1} + \rho_1 q_{\beta2})} \approx \frac{1}{2}\rho\alpha\sec\theta, \tag{C.16b}$$

$$\begin{aligned}\frac{\mathrm{F}^{\mathrm{N}}_{\mathrm{non_w}}}{\mathrm{A_w}}P^2 &= \frac{-4\rho_1\rho_2^2\beta_2^2 q_{\beta1}P^2(1-\beta_2^2 q_{\alpha2}q_{\beta2}) - 4\rho_1^2\rho_2\beta_1^2 q_{\beta2}P^2(1-\beta_1^2 q_{\alpha1}q_{\beta1})}{(\rho_2 q_{\alpha1} + \rho_1 q_{\alpha2})(\rho_2 q_{\beta1} + \rho_1 q_{\beta2})}P^2 \\ &\approx \frac{-2\mu}{q_\alpha}P^2 + 2\mu^2\frac{q_\beta}{\rho}P^2\frac{\Delta\rho}{\rho} - \mu^2\frac{q_\beta}{\rho}(1+\sin\theta^2)P^2\frac{\Delta\alpha}{\alpha} + 4\mu^2\frac{q_\beta}{\rho}P^2\frac{\Delta\beta}{\beta}.\end{aligned} \tag{C.16c}$$

Thus,

$$\begin{aligned}\frac{N^{\mathrm{N}}_{\mathrm{non_w}}}{D_{\mathrm{w}}} &\approx \frac{1}{2}\rho\alpha\sec\theta - \frac{2\mu}{q_\alpha}P^2 + \left(2\mu^2\frac{q_\beta}{\rho}P^2\right)\frac{\Delta\rho}{\rho} \\ &- \left(\mu^2\frac{q_\beta}{\rho}(1+\sin\theta^2)P^2\right)\frac{\Delta\alpha}{\alpha} + \left(4\mu^2\frac{q_\beta}{\rho}P^2\right)\frac{\Delta\beta}{\beta}.\end{aligned} \tag{C.17}$$

Therefore, the approximate PP reflection coefficient for the fractured medium finally is given by

$$
\begin{aligned}
\mathrm{R}(\theta) &\approx R_{\mathrm{w}} + \frac{i\omega S_{\mathrm{T}}}{1 - i\omega v_x S_{\mathrm{T}}} R^{\mathrm{T}}_{\mathrm{non_w}} + i\omega S_{\mathrm{N}} R^{\mathrm{N}}_{\mathrm{non_w}} \\
&\approx \left(2\frac{1}{\rho}\mu^2 q_\alpha P^2\right)\frac{i\omega S_{\mathrm{T}}}{1 - i\omega v_x S_{\mathrm{T}}} + \left(\frac{1}{2}\rho\alpha\sec\theta - \frac{2\mu}{q_\alpha}P^2\right) i\omega S_{\mathrm{N}} \\
&\quad + \left\{\frac{1}{2}\left(1 - 4\left(\frac{\beta}{\alpha}\right)^2 \sin^2\theta\right) + \left(\frac{\mu}{q_\beta}P^2 - \frac{1}{4}\rho\beta\sec\vartheta\right)\frac{i\omega S_{\mathrm{T}}}{1 - i\omega v_x S_{\mathrm{T}}} + \left(2\mu^2\frac{q_\beta}{\rho}P^2\right) i\omega S_{\mathrm{N}}\right\}\frac{\Delta\rho}{\rho} \\
&\quad + \left\{\frac{1}{2\cos^2\theta} + \left(\frac{\mu}{q_\beta}P^2 - \frac{1}{4}\rho\beta\sec\vartheta\sec\theta^2\right)\frac{i\omega S_{\mathrm{T}}}{1 - i\omega v_x S_{\mathrm{T}}} - \left(\mu^2\frac{q_\beta}{\rho}(1 + \sin\theta^2)P^2\right) i\omega S_{\mathrm{N}}\right\}\frac{\Delta\alpha}{\alpha} \\
&\quad + \left\{-4\left(\frac{\beta}{\alpha}\right)^2 \sin^2\theta + \left(2\frac{\mu}{q_\beta}P^2\right)\frac{i\omega S_{\mathrm{T}}}{1 - i\omega v_x S_{\mathrm{T}}} + \left(4\mu^2\frac{q_\beta}{\rho}P^2\right) i\omega S_{\mathrm{N}}\right\}\frac{\Delta\beta}{\beta}.
\end{aligned}
\tag{C.18}
$$

Index

© Springer Nature Singapore Pte Ltd. 2018

X. Cui et al., *Seismic Forward Modeling of Fractures and Fractured Medium Inversion*, Springer Geophysics, DOI 10.1007/978-981-10-3584-5

MIX
Papier aus verantwortungsvollen Quellen
Paper from responsible sources
FSC® C105338

If you have any concerns about our products,
you can contact us on
ProductSafety@springernature.com

In case Publisher is established outside the EU,
the EU authorized representative is:
Springer Nature Customer Service Center GmbH
Europaplatz 3, 69115 Heidelberg, Germany

Printed by Libri Plureos GmbH
in Hamburg, Germany